STEEL BARRIO

CULTURE, LABOR, HISTORY SERIES

General Editors: Daniel Bender and Kimberley L. Phillips

The Forests Gave Way before Them: The Impact of African Workers on the Anglo-American World, 1650–1850
Frederick C. Knight

Unknown Class: Undercover Investigations of American Work and Poverty from the Progressive Era to the Present
Mark Pittenger

Steel Barrio: The Great Mexican Migration to South Chicago, 1915–1940
Michael Innis-Jiménez

STEEL BARRIO

THE GREAT MEXICAN MIGRATION TO SOUTH CHICAGO, 1915–1940

MICHAEL INNIS-JIMÉNEZ

NEW YORK UNIVERSITY PRESS
New York and London

NEW YORK UNIVERSITY PRESS
New York and London
www.nyupress.org

LIBRARY OF CONGRESS CATALOGING-IN-PUBLICATION DATA

Innis-Jiménez, Michael.
Steel barrio: the great Mexican migration to South Chicago, 1915-1940 / Michael Innis-Jiménez.
pages cm. — (Culture, labor, history series)
Includes bibliographical references and index.
ISBN 978-0-8147-8585-0 (cl : alk. paper) — ISBN 978-0-8147-2465-1 (pb : alk. paper)
1. Mexican Americans—Illinois—Chicago—History—20th century. 2. Immigrants—Illinois—Chicago—Social conditions—20th century. 3. Working class—Illinois—Chicago—Social conditions—20th century. 4. Steel industry and trade—Illinois—Chicago—History—20th century. 5. Chicago (Ill.)—Emigration and immigration—History—20th century. 6. Mexico—Emigration and immigration—History—20th century. 7. South Chicago (Chicago, Ill.)—History—20th century. 8. Chicago (Ill.)—History—20th century. I. Title.
F548.9.M5I66 2013
305.89'6872077311—dc23
2012049430

New York University Press books are printed on acid-free paper, and their binding materials are chosen for strength and durability. We strive to use environmentally responsible suppliers and materials to the greatest extent possible in publishing our books.

Manufactured in the United States of America
c 10 9 8 7 6 5 4 3 2 1
p 10 9 8 7 6 5 4 3 2 1

To Heather M. Kopelson
For sharing this journey

In memory of Irene Ledesma
She inspired me to become an activist and historian

and

In memory of James M. Salem
His legacy lives on

CONTENTS

LIST OF ILLUSTRATIONS

ACKNOWLEDGMENTS

I first fell in love with the history of Mexican Chicago in Shelton Stromquist's urban history seminar. Since then, many people and organizations have contributed to this undertaking. I owe them a large debt of gratitude.

At the University of Iowa, Shelton Stromquist and Malcolm Rohrbough provided invaluable guidance for my work. I am also very thankful to Ned Bertz, Becky Pulju, Nat Godley, Shannon Fogg, Lionel Kimble, Jesse Spohnholz, John McKerley, Megan Threlkeld, Bob Bionaz, and Dana Quartana, for their support and friendship. While at Iowa, I was fortunate to receive the multi-year Jonathan W. Walton Fellowship and the Elizabeth Bennett Ink Fellowship. I also benefited from a Consortium for Institutional Cooperation (CIC) fellowship and from financial support provided by the King V. Hostic Award of the Illinois Historic Preservation Agency and the Illinois State Historical Society. Dean William Welburn was always there to help me patch together funding when I really needed it.

I also wish to recognize the University of Iowa graduate employees who fought for union recognition before I arrived there. Thanks to those who remain active in COGS/UE Local 896. The work of this member-run union has improved not only pay, benefits, and overwork protection, but also the physical and intellectual quality of life at Iowa.

I was fortunate to spend the 2003–2004 academic year writing in the beautiful city of Burlington, Vermont with the support of the George Washington Henderson Scholar-in-Residence program. I am grateful to the members of the University of Vermont's History Department for their sponsorship. They were wonderful hosts and mentors. I especially want to recognize Denise Youngblood, Willi Coleman, Melanie Gustafson, Kathy Carolin, and Kathleen Truax. I appreciate the time and friendship of Peg Boyle Single. This fellowship was part of the Northeast Consortium for Faculty Diversity directed by Joanne Moody.

In the fall of 2004, when I began teaching at William Paterson University of New Jersey and working further on this book, several colleagues were especially generous with their time and support, including History Department Chairs Terry Finnegan and Evelyn Gonzalez and Dean Isabel Tirado. Many thanks are also due to Lucia McMahon, Malissa Williams, Jason Ambroise, and the late Ana Gomez. They made WPUNJ a fun place to teach and write. While at William Paterson, I received financial support from the Research Center for the Humanities and Social Sciences Summer Research Stipend, the Assigned Release Time program, the Travel Incentive Grant program, and the Career Development program.

I have been in the American Studies Department of the University of Alabama since the fall of 2008. My colleagues in Tuscaloosa are wonderful. I will forever owe my department chair, Lynne Adrian, a debt of gratitude. She has been a supporter and advocate since my first visit to campus. I also appreciate the support of Rich Megraw, Ed Tang, Stacy Morgan, Jolene Hubbs, Eric Weisbard, Jeff Melton, and Ellen Spears. The Department of American Studies and the College of Arts and Sciences have supported research related travel. I have also benefited from a grant from the University of Alabama's Research Grant Committee.

In Chicago, Dominic Pacyga's suggestions and contacts were invaluable. Peter Alter's guidance and recommendation that I listen to the Jesse Escalante Oral History Collection changed the direction of my research. Many of the historical pictures are courtesy of the Southeastern Chicago Historical Society. I am grateful to Rod Sellers for his patience and guidance as I decided which photos to include. I also value the time and energy of Chicago photographer Curtis Myers who has provided me with many pictures of early twenty-first-century South Chicago. Thank you also to Christopher Winters, map bibliographer at the University of Chicago Library, for assisting with the old maps.

I also appreciate the contributions of the following people: Margaret Abruzzo, Jose Alamillo, Luis Alvarez, Mike Amezcua, Francis Aparicio, Bonnie Applebeet, David Badillo, Emma Bertolet, Gerry Cadava, Eduardo Contreras, Cary Cordova, Arlene Davila, Lilia Fernández, Leon Fink, Christina Frantom, Michel Gobat, Heather Godley, John McKiernan-Gonzalez, Colin Gordon, David Gutiérrez, Dennis Hidalgo, Jorge Iber, Rita Arias Jirasek, Malachy McCarthy, Susan Oboler, Jon Payne, John Raeburn, Carlos Tortolero, Wilson Valentin, Julie Weise, John Williams-Searle, and Carmen Whalen.

I have presented parts of this manuscript, in various stages of completion, at seminars and conferences. Thank you to everyone who has commented on the papers and presentations. I especially appreciate the work of the organizers of The Newberry Seminar in Borderlands and Latino Studies, The Chicago History Museum's Urban History Seminar, The Newberry Library's Labor History Seminar, and the 2012 Mexican Chicago symposium sponsored by The Latina and Latino Studies Program at Northwestern University. Stephen Pitti has been a supporter, a reader, and an advocate throughout the years.

I have met several people at conferences and at the archives that have supported this project from its earliest stages. I first met Gabriela Arredondo at what was then the Chicago Historical Society while I was doing preliminary research and she was wrapping up her book. She invited me to breakfast the next morning. She has been an incredible mentor ever since. Thank you for your confidence and time. I also very much appreciate Matt García's mentorship and counseling. Shortly after I met Gabriela, she introduced me to Matt at a conference. Our conversations about this project were always exciting and motivating. Vicki Ruiz and Ramon Gutierrez have encouraged me through the years. Thank you.

Steel Barrio would not be in your hands right now without the energy, encouragement, and support of my editor at NYU Press, Deborah Gershenowitz. Debbie has been a wonderful guide through the publication process. Constance Grady, also at NYU Press, has cheerfully put up with dozens of my questions. I also value the input by two anonymous readers. Their encouragement and suggestions improved this manuscript. Cleo and Carla Thomas provided funding that helped make this book possible. Thank you.

My parents, Art and Myrna Innis, have believed in me and supported my adventures. My children have been truly wonderful. Brittany and Emily have gone from grade school to college during the life of this book. Mateo and Alessandra have inspired me in the final stages of *Steel Barrio.*

One other person has been a part of the project since the start. This is Heather Miyano Kopelson. She has listened, commented, edited, coached, motivated, inspired, and loved me throughout this endeavor—and all while working on her own book. I am forever grateful to her for sharing her life with me.

* * *

Earlier versions of chapter 8 were published elsewhere: "Organizing for Fun: Recreation and Community Formation in the Mexican Community of South Chicago in the 1920s and 1930s," *Journal of the Illinois State Historical Society* 98, no. 3 (Autumn 2005): 144–161; "Beyond the Baseball Diamond and Basketball Court: Organized Leisure in Interwar Mexican Chicago," *The International Journal of the History of Sport* 26, no. 7 (June 2009): 906–923. The latter article will soon be reprinted by Texas Tech University Press in *More Than Just Peloteros: Latino/a Athletes in U.S. Sports History,* edited by Jorge Iber.

INTRODUCTION

This book is about first-wave immigrants to a new area. It is about those who left danger for opportunity despite uncertainty. It is about individual people who worked, lived, died, played, and organized in and around the Chicago neighborhood called South Chicago. It is about the individual people coming together by creating clubs, societies, and teams to advocate, socialize, play, and endure despite harassment and discrimination. This story of early Mexican immigration to South Chicago is as relevant today as it was nearly a century ago.[1]

On March 10, 2006, ninety years after the first significant Mexican immigration to the Windy City, somewhere between 100,000 and 300,000 Latino immigrants and their supporters took to the streets to protest the Sensenbrenner Bill.[2] The U.S. House of Representatives had passed the bill three months earlier. The Sensenbrenner Bill made it a felony for undocumented immigrants to live in the United States or for anyone to "aid, abet," or "counsel" an undocumented immigrant. Immigrant advocates credit this Chicago protest march as the event that galvanized a nationwide protest movement, taking place a full two weeks before 500,000 Latinos and their supporters rallied in Los Angeles. Chicago's largely Mexican immigrant population, with the support of Mexican Americans and other immigrant allies, reacted against a bill that targeted non-white immigrants. The first organizing meeting was held February 15, at headquarters of Casa Michoacán.

Today, Casa Michoacán is one of the most prominent Chicago-based Mexican hometown associations. Working much like mutual aid societies, these hometown associations were originally informal groups created by immigrants to "function as social networks as well as transmitters of culture and values to the U.S.-born generation."[3] Many associations that matured around the turn of the twenty-first-century have focused on larger, more politicized goals. Casa Michoacán and affiliated Chicago-based *Michoacáno* hometown associations have arguably led the way in advocating for social development projects back home in their communities of origin as well as in actively defending Mexican immigrant rights in the Chicago area.[4]

Like the Mexican immigrants who sought support from the Mexican consul against harassment and discrimination during the interwar years, Chicago Mexican activists during the 2006 protests requested support from Mexico

Fig. I.1. Pro-immigration rally participants march through the canyons of the Chicago Loop, March 10, 2006. Photograph by Joseph Voves.

City. Emma Lozano and Artemio Arreola, Mexican community leaders and organizers for the March 10 protest, were part of a Chicago delegation that visited Mexico City shortly after the protest to seek support from the Mexican government. At the Mexican presidential residence, Chicago community leader and immigrants' rights activist Juan Salgado argued that Chicago was "recognized as the place where all of these marches were born."[5] Lozano, Salgado, Arreola, and other protesters raised issues that have persisted throughout much of United States history, especially those pertaining to: racialization of particular immigrant ethnic groups and the targeting of these groups as distinct and somehow dangerous to the "American" way of life.

How did Chicago, an area with virtually no Mexican presence before 1916, become a national hub of Mexican and Mexican-American activism where, by 2010, 557,000 people of Mexican descent lived in the city and comprised almost 22 percent of its population?[6] Why did Mexican Americans feel a close bond to Mexican immigrants in a large city of immigrants and traditional Anglo-American power? To understand how the Chicago area became a hub for Mexican Americans and Mexican immigrants, we should examine closely the origins of the community and the factors that led to a vibrant and continually growing Mexican Chicago. To do this, I have singled out the early years of a Mexican community in one industrial neighborhood within the city and its relationship with other communities. South Chicagoan Jesse Escalante's gathering of oral histories in the early 1980s and his subsequent donation of his collection to the Chicago History Museum made it possible for me to take a close look at early Mexican South Chicago through the eyes of several community members.

Gilbert Martínez was one of these Mexican South Chicagoans. In 1980, Martínez sat down with Jesse Escalante to record a conversation about life there in the early years of the twentieth century. Martínez arrived in South Chicago in 1919 at the age of nine and would become a steelworker. Escalante was born in the neighborhood in 1924 and would become a civil servant and community leader. Very early on in the interview, Martínez turned to one of his favorite topics and what had been one of his favorite pastimes just before and during the Great Depression. "So what we used to do is put on the uniform, with spikes and everything and go over there and get the number 5 streetcar on Sundays" reminisced Martínez. "We used to take out bats and balls—a couple of new balls—and go to Washington Park."[7] In the audio recording, one can hear the pride and joy in Martínez's voice as he reminisced about being part of the Mexican community of South Chicago as a teenager and young man in the 1920s and early 1930s. Before long, Martínez and his teammates were "tangling up, playing ball with the Irish." Getting on a streetcar and traveling the 6 1/2 miles to the large park in the neighborhood of the same name is significant on several levels. First, most contact between Mexican youth and youth of other ethnic groups within their own neighborhood of South Chicago was usually negative and led to confrontation. The neighborhood's Polish and Irish residents made it difficult for Mexicans to use South Chicago parks. In order to gain regular access to baseball fields and avoid trouble, Martínez and his teammates felt that they had to leave their regular stomping grounds and go where anti-Mexican tensions were less pronounced. Second, Martínez and his teammates set out to represent themselves and their neighborhood in a positive way by putting on clean uniforms and using the best equipment they had. Third, baseball became an avenue for the new and exciting as players looked for opportunities to play in other parts of town or in nearby Northwest Indiana.

Similarly to new immigrant populations in the United States today, Mexicans in South Chicago dealt with economic hardship, ethnic prejudice, nativism, and intra-ethnic divisions. These factors reinforced their sense of difference and their propensity to see themselves as sojourners desiring to return to Mexico as soon as possible. However, this sojourner attitude was not an absolute obstacle to the creation and support of a Mexican culture in South Chicago. Mexicans who migrated to South Chicago—in a pattern that began with workers being hired by labor agents along the Texas-Mexican border in 1919—entered a neighborhood that was already blighted by decades of environmental racism and were confined to clearly demarcated community and workplace niches. They came through the encouragement of friends, after being recruited in Mexico or along the border, or after many years of working in other parts of the United States Midwest and West.

Until recently, the assumption of most scholarship on Mexicans in the Chicago area was that weak, fractured communities existed within the three major concentrations of Mexicans in the city and that little area-wide cooperation existed because Mexicans tended to stay, work, and play, within their respective neighborhoods. South Chicago Mexicans, however, did create links and cross-community organizations in order to improve their environment and defend against the social, political, and economic harassment and discrimination that plagued their everyday lives. These ties to Mexicans in the Near West Side and Back-of-the-Yards neighborhoods of Chicago, as well as in Gary and East Chicago, Indiana, helped South Chicago Mexicans survive the Great Depression as a distinct community. They might have used survival skills they learned from wartime shortages and uncertainty in places like Guanajuato, Michoacán, and Jalisco, or they might have learned the survival skills they perfected after years of mistreatment and neglect as immigrant workers in the United States. For many, a combination of these experiences hardened them and helped them survive. They also survived and persisted with help from community leaders, social workers, immigrant advocates, and others in South Chicago. And they found further strength through a common cultural bond to and identification with Mexico.

Individuals such as steelworker Justino Cordero, activist and social worker Mercedes Rios, grocery store owner and Yaquis sports club founder Eduardo Peralta, and steelworker and union organizer Alfredo De Avila emerged from Mexican South Chicago as leaders. They, along with others, formed a community that was able to change its physical and cultural environment to help its members and create a degree of resistance that enabled Mexicans to persevere against the intimidation and prejudice that rose exponentially as the national and local economy faltered.

The idea that such leaders emerged might not seem exceptional or new, but scholars have not paid much attention to this development. Contemporary scholars who studied Mexicans in Chicago before World War II assumed that such leadership did not exist—or need to exist—because of a strong Mexican consular presence, Catholic and Protestant organizations, and settlement houses. The settlement house work with Mexicans included teaching them how to negotiate local and federal government mazes and protecting them from obvious forms of discrimination and harassment by property owners, employers, and city leaders.

Although settlement house workers did play important roles in the Near West Side and Back-of-the-Yards Mexican communities, the lack of a large, secular settlement house in South Chicago placed a larger burden on South Chicago community members to lead from within. This need for Mexican leadership became most significant in times of economic crisis, when white and ethnic European Chicagoans blamed *Mexicanos* for lower wages and lack of jobs.

The acute crisis of the Great Depression and the subsequent movement of *Mexicanos* out of the neighborhood caused profound changes in the Mexican community of South Chicago. First, there was significant, mostly voluntary, outmigration to Mexico of members of the community. Second and simultaneously, those who stayed organized themselves through political, religious, social, recreational, and *pro-patria* groups and claimed their rights as residents of the United States. These factors helped in the creation of a post-depression Mexican community in South Chicago which was much smaller than before, but was better able to fend for itself and to organize against economic, social, and political discrimination and harassment. Focusing on these trends, this study ends in 1940, when the economic rebound caused by

World War II in Europe brought a renewed demand for new immigrant labor from Mexico and the beginning of a new era for Mexican South Chicago. By 1940, leaders in a much smaller and highly organized community sought to distance themselves from the new immigrant generation, which looked much like the one that entered the Chicago area in 1916 and South Chicago in 1919.

This study goes beyond the workplace. As with most migration, work was the primary reason the vast majority of Mexicans came to South Chicago. The large industrial buildings where these migrants worked—the steel mills dotting the landscape—helped define the neighborhood. Although I take into consideration the workplace and the unions that fought to organize the workers, I am primarily concerned with the lives of working-class Mexicans in an urban, industrial, ethnic neighborhood. They organized independently of their workplaces in *mutualistas,* social clubs, and church organizations. They resisted harassment by government officials, bosses, coworkers, union organizers, property owners, and others who racialized their existence in Chicago. This study is, then, also an examination of how Mexicans persisted in the steel barrio despite the steel mills. Steel mills provided jobs for Mexican men, but were not at the center of community life or the source of collective identity. Discrimination by company management and labor unions prevented Mexicans from identifying closely with the workplace or the union hall. Surviving thus meant creating community outside the mill gates, but in a neighborhood whose existence depended on steel mills.

There is a vast literature on Chicago as an urban center in the Midwest, on various ethnic communities in Chicago, and on the labor histories of meatpacking and steel.[8] However, this study of an emerging Mexican community in the Midwest is most directly in conversation with scholarship focused on urban Mexicans in the United States. Within this field, the study of the history of Mexicans in the industrial Midwest is a relatively new development and that of Mexicans in the Chicago area an even more recent one..

Paul S. Taylor and Manuel Gamio completed sociological and anthropological studies of Mexicans in the United States that included significant research on the Chicago area in the late 1920s and early 1930s. Arguably the two most important archival collections for the study of the Mexican population of interwar Chicago, their collections are housed at the Bancroft Library at the University of California, Berkeley. Not only did both scholars conduct extensive interviews of Mexicans in the Chicago area and throughout the country during this period, they also recorded their own observations relevant to the community and interviewed employers and officials who were in contact with Mexicans in the community. Both of these collections are invaluable to the study of Mexican Chicago and provide the vast majority of first-person accounts and interviews of and about Mexicans in Chicago during the interwar years. Gamio's books resulting from his study included *The Mexican Immigrant: His Life Story* and *Mexican Immigration to the United States: A Study of Human Immigration and Adjustment*.[9] Taylor's volume on Mexicans in the Chicago area, *Mexican Labor in the United States: Chicago and the Calumet Region*, is part of a multi-volume series on Mexicans in the United States. Also notable are the vast majority of extant Spanish-language Chicago newspapers collected and saved by Manuel Gamio.[10]

In the same period, students at the University of Chicago School of Social Work produced several relevant studies. Other Chicago-area scholars produced useful journal articles, WPA reports, and church sponsored publications. While many were merely paternalistic overviews of Mexicans in the area, a few stand apart. Most significant of these is Robert C. Jones and Louis R. Wilson's report titled, *The Mexican in Chicago, The Racial and Nationality Groups of Chicago: Their Religious Faiths and Conditions*. Later in the book, I will discuss this report in detail.[11]

The next significant scholarly work on Mexicans in the industrial Midwest did not come out until the 1970s, when it appeared in conjunction with and as part of the nationalist Chicano movement. Most of it concentrated either on the entire urban Midwest, on Detroit, or on Northwest Indiana.[12] The only substantial scholarly work to focus exclusively on Mexicans in Chicago was Louise Año Nuevo-Kerr's 1976 "'The Chicano Experience in Chicago, 1920–1970,"[13] a groundbreaking study of urban Mexican communities outside of Texas and California which focuses primarily on the "establishment and evolving differentiation of the Chicano settlements in Chicago" during four distinct periods between 1916 and 1970. In her study, Año Nuevo-Kerr argues that because Mexican communities existed in separate neighborhoods of Chicago, these disparate communities did not evolve in the same manner. Thus, Mexicans in these communities experienced "a differential development of conscious ethnicity and community identity."[14] After Año Nuevo-Kerr, Gabriela Arredondo produced pioneering work in the early 2000s. In *Mexican Chicago: Race, Ethnicity and Nation, 1916–1939*, she focuses on the racialization of Mexicans city wide and the establishment of a fragile *Mexicanidad* throughout the city. *Steel Barrio* builds on Arredondo's work on racialization and *Mexicanidad* to explore how and why Mexicans in South Chicago were able to use their physical and cultural environment to develop organizations that started as sports and social clubs, and then became anchors in the community's determination to survive and persist.[15] I also build on Arredondo's discussion of cross-community links throughout the area.

In addition, Lilia Fernández's 2012 monograph, *Brown in the Windy City: Mexicans and Puerto Ricans in Postwar Chicago,* is an extremely significant contribution to the study of the racialization of post–World War II Mexicans in the City of Chicago. Fernández comparatively examines the racialized Mexican and Puerto Rican struggle for "place" in a city that racialized them in an inconsistent and fluid manner throughout the second-half of the twentieth century. Further important scholarship includes John Henry Flores's "On the Wings of the Revolution: Transnational Politics and the Making of Mexican American Identities," which examines the development of political culture in Mexican Chicago, its links to political developments in Mexico, and how both shaped identity for Mexicans in Chicago. Mike Amezcua's "The Second City Anew: Mexicans, Urban Culture, and Migration in the Transformation of Chicago, 1940–1965," discusses the significance of cultural sites and sites of economic development as crafters of shifting ethnic politics and identity post-World War II Mexican Chicago.[16]

The first works published in the wake of the Chicano movement argued against the once dominant "ghetto model" of interpreting Mexican urban communities, which emphasized the negative aspects of the urban Mexican community life and characterized their neighborhoods as sources of an array of criminal activities. These works contributed to the development of the field by looking at urban areas as more than mere incubators for delinquency. Scholars argued that the ghetto model removed any agency from the Mexicans in their communities and perpetuated the idea of Mexicans' victimization. The vast majority of these community studies focused on Southern California, with Texas a distant runner-up.

Historians argue that the interwar period, primarily the period surrounding the Great Depression, served as the critical period in the formation of a new ethnic Mexican-American identity in the United States. The economic hardships, repatriation campaigns, and intense harassment of Mexicans shaped a new cultural identity that worked in opposition to—or as resistance against—pressures to Americanize or repatriate. Mexicans in the United States viewed the retention of *Mexicanidad,* or a cultural Mexicanness, within an adapted ethnic Mexican cultural identity throughout and after the Great Depression as a crucial component to survival and resistance against pressures from the dominant society.[17]

The studies on outside effects on Mexicans in the United States tend to converge around the repatriation movements of the 1930s and 1950s and the post–World War II political, social, and educational forces that worked against the population. The organized repatriation movement during the Great Depression destabilized Mexican communities throughout the United

States and in Mexico. Mexicans in areas such as Los Angeles, Detroit, or Gary, Indiana, regardless of citizenship or immigration status, lived in fear of forced deportation of themselves or family members.[18] Mexicans in Chicago did not experience a significant, organized forced repatriation drive, but they did live in fear of harassment and deportation.[19]

A significant group of studies analyzes class tensions within Mexican communities. Among these are monographs that examine contributions to culture and community made by middle-class Mexican Americans, who did not experience the extent of the discriminatory practices that weighed heavily on new immigrants and working-class Mexicans.[20] Because of the newness of Mexicans communities in Chicago, and because the vast majority of Mexican immigrants came to Chicago to work in low- or unskilled labor, no discernible "Mexican-American" middle class existed in the area during the 1920s and 1930s. Such was also the case in other Midwestern cities such as Detroit, Kansas City, and St. Paul. Although a very small Mexican intelligentsia did exist in Chicago, they maintained an elite Mexican identity, which provided them with a status that kept them from having to negotiate with the dominant society to avoid harassment or discrimination.[21]

Zaragosa Vargas, in *Proletarians of the North: A History of Mexican Industrial Workers in Detroit and the Midwest, 1917–1933*, draws a complex portrait of post–World War I Mexican communities, job mobility, and the influx of workers who were lured away from seasonal railroad and agricultural jobs for high-paying ($5 a day) jobs at the Ford plant. As in other cities, Mexican immigrants in Chicago and Detroit were active agents within their community who built mutual aid societies, clubs, and fraternal organizations in order to function and adapt to the needs of the community while maintaining a cultural link to Mexico. Unlike the other scholars, Vargas uses a more worker-centered approach to argue that Mexican male workers adapted to an American industrial proletariat work ethic despite racial discrimination. Because of this, Vargas argues, members of the community were ambivalent with regard to the racial hierarchy that placed Anglo Americans and European immigrants above Mexicans while placing African-American workers below them in both the workplace and society. By contrast, I argue that Mexicans in Chicago were not ambivalent about their racialization and that much of their resistance was centered on this race-based harassment and discrimination.

More recent books on Mexicans in the Midwest more closely analyze culture and gender within the community. In *Barrios Norteños: St. Paul and Midwestern Mexican Communities in the Twentieth Century*, Dionicio Nodín Valdés acknowledges the important role gendered internal and external power relationships had on the everyday lives of *Mexicanos* in Mexico and the Midwest. He examines how the dominant society affected power relationships, interpersonal relationships, and socially constructed roles within the Mexican community.[22] Outside of work produced by a handful of authors who focus directly on urban and rural Mexican women, very few studies focusing on gender in Mexican communities exist. One study that does describe the ability of a community to survive while focusing on gender is Sara Deutsch's investigation of Mexican society in northern New Mexico and Colorado. Although the Mexican community there predates Anglo-American settlement in the area, *No Separate Refuge* is useful for the study of the Mexican community of Chicago because it provides an example of how a community that has been commonly characterized as "isolated, static, inflexible, paternalistic, and passive" was able to adjust.[23] While Vicki Ruiz and Julia Blackwelder have focused on women's experiences in a multi-ethnic/multi-racial workplace,[24] Donna Gabaccia's *From the Other Side* is a study of women and gender in immigrant communities that starts by looking at the experiences of immigrants before they leave for the United States and then examines the changes that occur inside and outside of the household once they are in the United States. She addresses migration, labor, family, class, and community activism while comparing the histories of migrants with those of Anglo and African Americans to understand how they construct their identities.[25] Ruiz's *Out of the Shadows* examines the conflicts between Mexican parents and their daughters in the Southwest caused by the girls' "Americanization," particularly in social activities.[26]

Steel Barrio explores several themes within the community and community members' interactions with the environment. Community itself is one of these themes. Having a clear understanding of the concepts of community and a sense of belonging to a community is important to effectively examine how Mexicans in South Chicago came together to change their environment. When people share a culture, resources, and the use of physical spaces in a single geographical location, they form community not only through interacting with one another but also by considering themselves part of the group. In other words, people create a community when they have a sense of being in an environment where they share common experiences, norms, and cultural understanding; they must also believe that they share a common bond and concern for one another. That being said, communities are dynamic, and not all members will share all elements of the community.[27] David Gutiérrez defined the zone of safety Mexicans in California tried to make for themselves to ameliorate their experiences of dislocation and discrimination as a "third space" within ethnic enclaves, whether urban barrios or rural *colonias*. "Located in the interstices between the dominant national and cultural systems of both the United States and Mexico," this "third" social space was where Mexicans "attempted to mediate the profound sense of displacement" and other pressures of being "members of a racialized and marginalized minority."[28]

Steel Barrio focuses on Mexicans in South Chicago and examines the details of the third space that they created for themselves and with other Mexicans in the area. This third space in South Chicago included sites for celebrations, sports venues, mutual aid societies, and other alternative arenas where Mexicans could work against the effects of discrimination. I include celebrations in this space of resistance because when members of the community celebrated Mexican cultural holidays, they visibly represented and enacted their Mexican culture, looking to hold off the pressure to blend seamlessly into mainstream society. The assimilationist idea that Mexicans should seamlessly blend into the dominant society was unrealistic since Mexicans were racialized as less-than-white and therefore unable to become part of the dominant "white" society.

In examining how external pressures on Mexicans in South Chicago to assimilate, along with constant harassment and discrimination, played prominent roles in how Mexicans shaped their environment, their community, and a distinct culture, my work builds on studies of Los Angeles by early scholars who came out of the core of the Chicano movement. Among these, Ricardo Romo's examination of the effects that a rapid economic boom and extensive Mexican migration had on the city's Mexican community, as well as his study of how Americanization programs put significant social and cultural pressure on the Mexican enclaves, are useful as models for delineating the change over time in South Chicago's Mexican community.[29] Douglas Monroy's attention to the ways in which the isolation of the Los Angeles Mexican population helped develop a culturally and socially independent community, a *Mexico de afuera* (Mexico on the outside), demonstrates that prejudice and discrimination from the dominant society do not always have to lead to cultural assimilation.[30]

I also focus on the expected and actual processes of assimilation for Mexicans in South Chicago. Whiteness required assimilation. Assimilation was crucial in order for immigrants to become productive members of society—that was, at least, the predominant attitude of local and federal governments, immigrant advocacy groups, and anti-immigrant forces. Nativists viewed assimilation as a way to absorb a threatening population. The more liberal view within the dominant culture regarded assimilation as democratic and egalitarian since Americanized Mexican immigrants had greater employment and social opportunities. For nativists, the only alternative to assimilation was deportation; for immigrant advocates, the lack of assimilation would lead to a segregated society where those who did not assimilate would remain at the bottom of the social ladder.[31] For the last thirty-five years, Chicano scholars have debated the reasons for the maintenance of distinct Mexican cultures in the shadow of Americanization programs. Richard Griswold del Castillo endorses a nationalist position that emphasizes that the maintenance of Mexican culture was not a conscious act of resistance but a result of the barrioization of the community. In other words, Mexican culture thrived because of the creation of Mexican barrios and not because of external pressures. George Sánchez argues that early Chicano historians focused on the constraints on assimilation instead of concentrating on the "symbolic and transformative significance of culture."[32]

As a study of an emerging urban Mexican community in the United States, this book explores the techniques and strategies Mexicans used selectively to resist assimilationists' efforts to eliminate Mexican cultural practices and celebration, food, and the use of Spanish. Arnoldo de León describes the Mexican American middle-class community of Houston as a group that, beginning in the 1930s, "lived in a world where they could voluntarily select from '*lo americano*' and '*lo mexicano*,'" easily transitioning from one identity to the other whenever they perceived it to be in their best interest.[33] In *Becoming Mexican American*, George J. Sánchez explores the role continuous immigration to the Los Angeles area played in the type of assimilation Mexicans experienced in culture and politics and the extent to which these immigrants became "assimilated" Mexican Americans. Starting with the political economy in Mexico that pushed the immigrant out, and following through to the "emergent ethnicity" of the Mexican immigrant in the United States, Sánchez combines a new borderlands framework with a cultural and social study of Mexican immigrants and Mexican Americans. By doing this, Sánchez defines an evolving space inhabited by a heterogeneous community of Mexicans, dispelling the dichotomous notion of independent "Mexican" or "American" cultures. Sánchez argues that Mexicans frequently redefined themselves through resistance and acculturation in relation to changing conditions and daily experiences in and around their communities.[34]

Steel Barrio is divided into three parts. The first examines the reasons for and routes related to Mexican migration to South Chicago starting in 1916. Work, *enganchistas* (labor agents), revolution, riots, and legislation all contributed to the significant influx of Mexicans that started with the railroads in 1916 and expanded to the steel mills by 1919. Once the initial wave of immigrants established themselves in South Chicago, new migrants used chain and circular migration patterns to grow the community physically and culturally.

In the second part I argue that Mexicans who came to South Chicago confined themselves to clearly demarcated everyday-life and workplace niches that aided the creation of a community where members changed their environment to help those in the community survive, persevere, and sometimes thrive. They entered a preexisting, ethnic, working-class neighborhood where Mexicans were the newest in a seemingly endless stream of ethnic immigrants and African-American migrants from the American South. I examine the physical and cultural sites where a distinctive shared identity is formed—the neighborhood and the community. Two factors that set the Mexican immigrants of South Chicago apart from previous immigrant groups are: 1) the proximity of Mexico to the United States, which allowed for a circular migration that culturally refreshed the community through the contact many immigrants kept with those in their sending communities; and 2) the view held by most Mexicans in South Chicago that they were sojourners who would eventually return to Mexico financially secure and able to provide a comfortable life for themselves and their family. Most in the Mexican community did not see this sojourner attitude as making them a drain on the United States. After all, they were contributing to the industrial economy at a time when steel mills were in desperate need of workers.

The last part focuses on everyday Mexican life in South Chicago during the Great Depression. I argue that distinct events during the crisis in the Chicago area were critical in the evolution of a strong Mexican community. High unemployment, repatriation, and the development of organized sports contributed to a more physically and culturally entrenched community that endured. Although involuntary repatriation programs were not as prevalent in Chicago as compared to other cities, depression-era unemployment affected South Chicago Mexicans at least to the same degree that it affected Mexicans in other parts of the United States. I also examine how South Chicago Mexicans used recreation, primarily organized sports, not only to persist and persevere, but as a vehicle to create organizations that promoted positive cultural and physical environments that in turn improved everyday life for members of the community.

This book is about much more than baseball or one single type of organization. It is about how community members used the discrimination against them, a sojourner attitude, organized sports, mutual aid organizations, and other groups to bring together a diverse community of Mexicans and Mexican Americans—some educated, most not—to find ways to change their physical and cultural environment in order to survive. I examine how the fortunes of Mexicans in South Chicago were linked to the built environment, their access to green space, and to their ability to change their physical and cultural surroundings. While examining how and why Mexicans acted on the industrial landscape by creating physical and cultural communities, I link their use of the urban environment to their ability to create, survive, and at times thrive. The South Chicago empty lots, baseball fields, and parks that Gilbert Martínez and his teammates preferred to use instead of having to travel out of the neighborhood must be juxtaposed with the industrial furnaces and chimneys that dominated the landscape between the homes and the shores of Lake Michigan. These chimneys spewed pollutants that invaded Mexicans' homes, their lungs, their eyes, and their very cells. Changing attitudes by government and business leaders toward the land and definitions of acceptable levels of pollution were sometimes couched in racialized terms. On the other hand, those resistant to stricter standards of cleanliness of the air, water, and land sometimes blamed governmental inactivity on the people living in the space and their unenlightened "traditions." Thus, this microstudy of a particular community in South Chicago is linked to broad issues in twentieth-century United States history, including urban growth and death, segregation within cities, and the relationship of urban immigrants to their natural and built environment. Through its focus on individual lives and the environment, this study of Mexican experiences—important in their own right—provides a lens to identify what happens to an urban neighborhood when industry, government, and social agencies battle for influence and control of a racialized immigrant workforce *and* their environment.

Scholars agree that labels are fraught with problems. We will not all agree on the proper term for a specific group of people or on the specific borders of a neighborhood or a section within a neighborhood. We can, however, stave off some confusion by defining our terms. In *Steel Barrio* the ones identifying those of Mexican descent are among the most important.

Thus, "Mexican" and "*Mexicano*" identify Mexican immigrants and Americans of Mexican descent who preferred to identify as Mexican. These were the most common terms used by the community's mostly working-class population, and here, they function interchangeably. When it becomes necessary to differentiate between those born in the United States and those born in Mexico, I use the term "Mexican American" to identify naturalized or U.S.-born Americans of Mexican descent. However, in this context "Mexican American" does not necessarily denote cultural assimilation into the dominant society. "Mexican immigrant" refers to those who immigrated to the United States from Mexico. The terms "Anglo" and "Anglo-American" describe non-Mexican, white residents of the United States. While Southern and Eastern European immigrants could sometimes be included in the "white" category, especially when the context emphasized Mexicans against all non-Mexicans, Mexicans in Chicago routinely distinguished between non-Mexicans they considered ethnic Europeans and those who were "American." In Chicago, using "Anglo" for the latter category was less common than it was in Texas or the Southwest, but it was a familiar term.

This book focuses on the neighborhood of South Chicago, an area officially incorporated into the City of Chicago and much further south than the area commonly referred to as Chicago's South Side. Mexicans in South Chicago were not isolated in their neighborhood and traveled to other parts of the city and to steel-mill towns in Northwest Indiana. I use the term "Chicago" when referring to the city as a whole or multiple city neighborhoods, and I use the term "Chicago-area" to include communities in Chicago and in Northwest Indiana. The cluster of neighborhoods that include South Chicago, East Side, South Deering, and Hegewisch is commonly referred to as the Southeast Side of Chicago and at times as the Far Southeast Side of Chicago. Mexican residents during the period of this study considered themselves part of the South Chicago Mexican community if they lived in any of these Southeast Side neighborhoods. For example, South Deering was home to Wisconsin Steel and a vibrant part of what many *Mexicanos* considered the South Chicago Mexican community. In addition, neighborhoods were further subdivided into areas such as the Bush and Millgate in South Chicago and Irondale in South Deering.

I

MIGRATION

1

Mexico and the United States

Steelworker and baseball player Gilbert Martínez moved to the neighborhood of South Chicago in 1926 at the age of sixteen. His journey to South Chicago started in 1914 when he and his family left Torreon, Coahuila, Mexico, in search of safety, work, and an escape from the Mexican Revolution. At the time, Torreon was the "railroad centre of the interior" where national and international rail lines connected. That made it an important strategic city, and one that was controlled by Pancho Villa's Northern Army for much of the revolution. Although the exact date of Martínez's departure from Torreon is unclear, major battles occurred in the area in April and December of 1914.[1]

Martínez's father had been working in "some plant," probably the large smelter just outside of town, when the fighting came through Torreon. Martínez, who was four years old when he left Torreon, does not talk much about the fighting, but he does remember the destruction. In a 1980 interview with the community leader Jesse Escalante, Martínez offered a succinct summary of the effects of the Mexican Revolution on hundreds of thousands: "Everything stopped and hardship." The simplicity of Martínez's statement belies the complexity of the universal devastation that caused everyday life and society to stop. People were threatened or killed for being on the "wrong side" in a revolution that had many armies and alliances. Work disappeared as the economy collapsed and people fled the chaos. The environmental devastation of war in the countryside severely limited food supply, adding an additional incentive to flee to those who were able. Although leaving Torreon was difficult, the Martínez family decided that the United States was their best option. Gilbert Martínez's extended family, "the whole clan," which included two of his uncles and their families, traveled to Cuidad Juarez in "those freight cars that carry sheep." They crossed into El Paso. There the women and children lived while the men shipped out for seasonal work on railroad crews, although railroad companies occasionally allowed the whole family to travel with the crews and live in boxcars. Three years after arriving in El Paso, the Martínez clan moved on. They contracted with an *enganchista*, or labor agent, for work on the railroads at various locations in the Midwest. Gilbert Martínez and his immediate family ended up in Galesburg, Illinois, while extended family members ended up in Iowa and Missouri. After six years of summer visits to his sister in Chicago, Martínez and his immediate family moved to South Chicago.[2]

In the twelve years of their journey from Torreon to South Chicago, the Martínez family experienced several facets common to those who left Mexico and found their way to South Chicago.[3] Although personal backgrounds and experiences varied greatly from person to person and family to family, generalizations are useful in examining overall migration patterns and illuminating how and why Mexicans entered this steel community. Tens of thousands—if not hundreds of thousands—left Mexico because of the economic difficulties created by the revolution. They hoped to find security, work, and a safe environment. Mexican men's entry into the United States industrial labor force was overwhelmingly, but not exclusively, through positions on railroad maintenance crews as *traqueros*, or track workers.[4] In fact, the first group of industrial, working-class Mexican men to enter the Chicago area came as *traqueros* under contract to various railroad companies extending or maintaining current railroad lines into the Chicago area. These contracts between Mexican workers and *enganchistas* on behalf of a railroad company commonly included transportation for the workers and their families to the worksite, usually charged to the new worker as an advance on wages. The first wave of Mexican migrants to the Chicago area started in 1916, a full decade before Gilbert Martínez's move to South Chicago.[5]

Although 1916 was the year that Mexican laborers first entered the Chicago area in any significant number, it was the 1919 steel strike and Chicago race riots, the Mexican Revolution and Cristero Rebellion, as well as legislation that restricted immigration from outside the Americas that, separately and together, created conditions that favored the migration of large numbers of Mexicans to Chicago and the rest of the industrial Midwest. After the first wave of Mexicans began to establish communities, the routes that Mexicans and Mexican immigrants used continued to broaden. One South Chicago settlement house leader believed that by 1928, only 25 percent of Mexican immigrants entering Chicago came directly from Mexico. He credited the additional migration to those quitting railroad jobs, those who "drift in from a variety of employments," and those who were "discontented farmworkers."[6]

That same year, Anita Edgar Jones divided the Mexicans coming into Chicago into five general classifications: those brought directly by industry; those who worked their way north in easy stages, passing from one job to another, or from one division of the railroads to another; those who came directly to Chicago upon the advice of others who were there or had been there; those who came via the sugar beet fields of Michigan, Wisconsin, or Northern Minnesota; and those who were *solos*, or unaccompanied men who had no ties to the United States and were "seeking adventure."[7]

According to the United States federal census, Mexicans comprised 1,141 of the 808,558 foreign-born residents in the entire city of Chicago in 1920.[8] Other sources that did not differentiate between Mexican immigrants and Mexican Americans listed the 1920 Mexican population of the city as 2,537. According to border records that indicate the intended destination of Mexican immigrants from 1920 to 1927, over 5,200 Mexicans listed Illinois as their final destination, with many of those undoubtedly bound for the Chicago area.[9] Because of the lack of reliable data as to the numbers of Mexican immigrants to the United States, it is difficult to ascertain the number of Mexicans who crossed the border into the United States knowing that Chicago was their final destination. Regardless of declared intentions, the official 1930 Mexican and Mexican-American population of Chicago had grown to 19,632. By 1940, the post–Great Depression Mexican community had shrunk to around 16,000.[10]

National numbers and percentages of European-born and Mexican-born immigrants demonstrate that South Chicago's increase in Mexican population was part of a nationwide shift in immigration patterns. Although a few Mexicans already lived in the area before 1916, that year marked a turning point in the creation of a significant Mexican presence in the Chicago area. The outbreak of World War I in Europe caused a virtual end to European immigration to the United States, as able-bodied men were either occupied with fighting or unable to leave because of the disruption of war.[11]

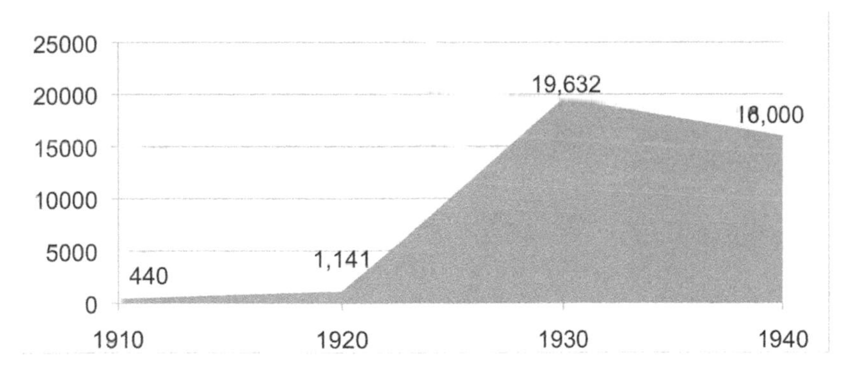

Fig. 1.1.Mexicans in Chicago, 1900–1940.

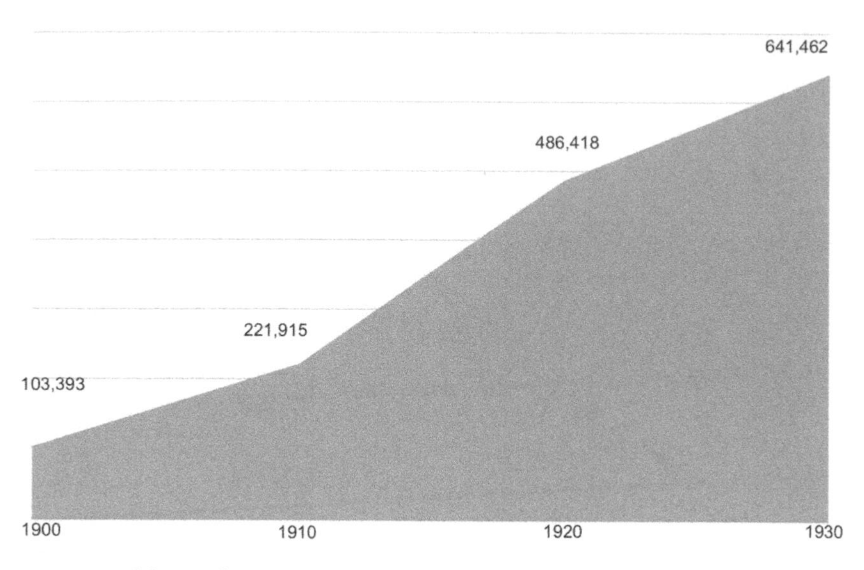

Fig. 1.2. Total foreign-born Mexicans in the United States, 1900–1930.

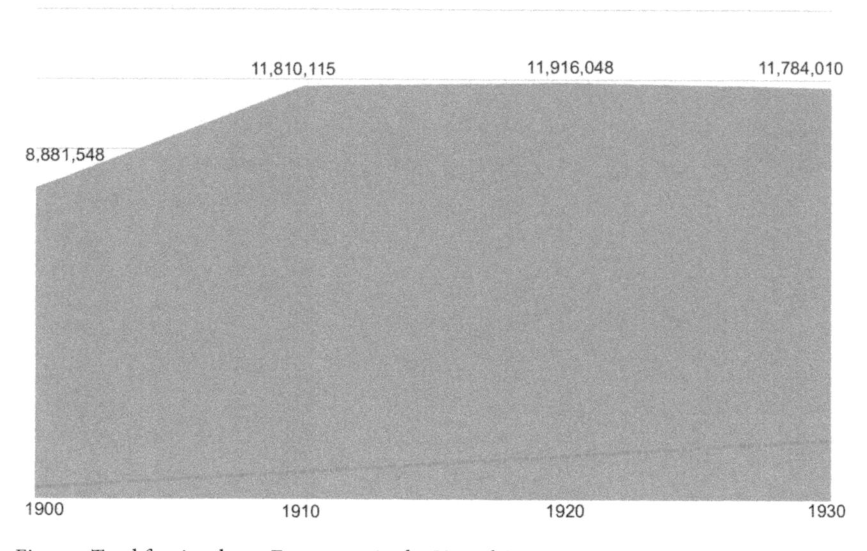

Fig. 1.3. Total foreign-born Europeans in the United States, 1900–1930.

In the first part of this study, I lay the foundation of the creation of a Mexican South Chicago by examining how and why Mexicans migrated there. I argue that various national and local factors shaped the patterns and timing of Mexican migration to the American Midwest in general and South Chicago specifically. In many cases, these factors directly affected the environmental conditions that awaited the newcomers along their journey and in South Chicago. The first chapter focuses on arguably the two largest, macro-level factors that shaped the timing and level of Mexican migration to the United States. The first of these was the revolutionary influence: the Mexican Revolution and the Cristero Rebellion prompted tens of thousands of Mexicans to migrate to the United States. The second factor was a result of restrictive U.S. immigration legislation that increased a labor shortage in the industrial Midwest: because Mexicans were exempt from these 1921 and 1924 quota laws, they became the default choice for employers who did not want to hire African Americans. In the following chapter, I focus on labor, labor agents, and racialized hiring practices. In the Chicago area, the demand for Mexican labor was accentuated by the race riots that immediately preceded the 1919 steel strikes. Racial tensions and the fears of some steel mill managers that hiring African Americans as strikebreakers would incite violence in South Chicago led to the recruitment and hiring of Mexicans. In short, revolution, riots, and race were significant factors in the migration of Mexicans to South Chicago.

Many early Mexican immigrants to South Chicago—those who came in 1916 through the 1920s—frequently cited the intertwined political, economic, social, and environmental repercussions of the Mexican Revolution as their impetus for leaving Mexico. Since the late nineteenth century, the United States has provided an escape-valve for Mexicans looking for opportunity and a possible break from economic despair.[12] The devastating effect of revolution on everyday Mexican life is arguably the most prominent factor prompting the large exodus of Mexicans to the United States.[13] War disrupted an already fragile agricultural economy, creating widespread environmental devastation, malnutrition, lack of work, and inflation. Many men volunteered to fight in one of the four principal armies of the Revolution of 1910, and many who did not want to fight fled to the United States fearing conscription. In addition, some fighters fled Mexico to avoid capture and conscription by advancing enemy armies. Civilians also feared for their lives as violence was not confined to the military battlefields.[14]

Although most of the fighting of the revolution ended by the time the United States entered World War I, continued unrest, including the Cristero Rebellion, caused turmoil in parts of Mexico throughout the 1920s.

The Cristero Rebellion of 1926–1929 cost 90,000 lives (35,000 Cristeros and 57,000 government soldiers and anti-Cristero forces), and was fought in the rural states of Jalisco, Michoacán, Durango, Guerrero, Colima, Nayarit, and Zacatecas. Not coincidently, these states comprised the core sending states for Mexican migrants to Chicago. The uprising primarily involved rural peasants and their local priests protesting the federal government's banning of the practice of Catholicism, including the celebration of mass and the administration of holy sacraments such as baptism and marriage.[15] The fighting and long-term political and economic instability led a steady stream of immigrants to move northward to South Chicago throughout the late 1910s and the 1920s.[16]

The dead often leave families behind. Soldiers and civilians killed during the violence of the revolution were no exception. Pancho Villa's army captured Serafín García's father shortly after the Mexican federal army had drafted him in 1914. In order to stay alive, García's father joined the Villista army and fought for two years. After returning to his job in a bakery, he was executed by soldiers of yet another revolutionary army, this time serving Venustiano Carranza. Serafín García, his mother, brothers, and some extended family fled the chaos of the Mexican Revolution in Guanajuato for Texas. They left a country at war and in social and economic disarray for the promise of a new life in the United States. After six years of farm work and odd jobs in Texas, and a season as *betabeleros,* or sugar-beet pickers, in Ohio, the García family settled in South Chicago.[17] García's mother left revolutionary Mexico with no intention of settling in the U.S. Midwest, but she, along with her children, eventually made their way to South Chicago and became active members of its Mexican community. García's family is an example of another route into the urban South Chicago Mexican community, through Midwestern farm work.

For many Mexican immigrants in South Chicago during the 1910s and 1920s, regardless of whether they migrated directly to South Chicago or moved there after first living in other parts of the United States, the economic chaos related to the Mexican Revolution was the leading macroeconomic factor that prompted many to consider migration.[18] A 1925 survey sponsored by the City of Chicago found that over half of the heads of households surveyed were from three central Mexican states: Guanajuato, Michoacán, and Jalisco. These three states were severely affected by fighting during the revolutionary period of 1910–1920, and bore the brunt of the violence during the Cristero Rebellion.[19] This demographic pattern held throughout the interwar period. By 1936, 63 percent of South Chicago Mexicans came from these same three states. The remaining 37 percent came

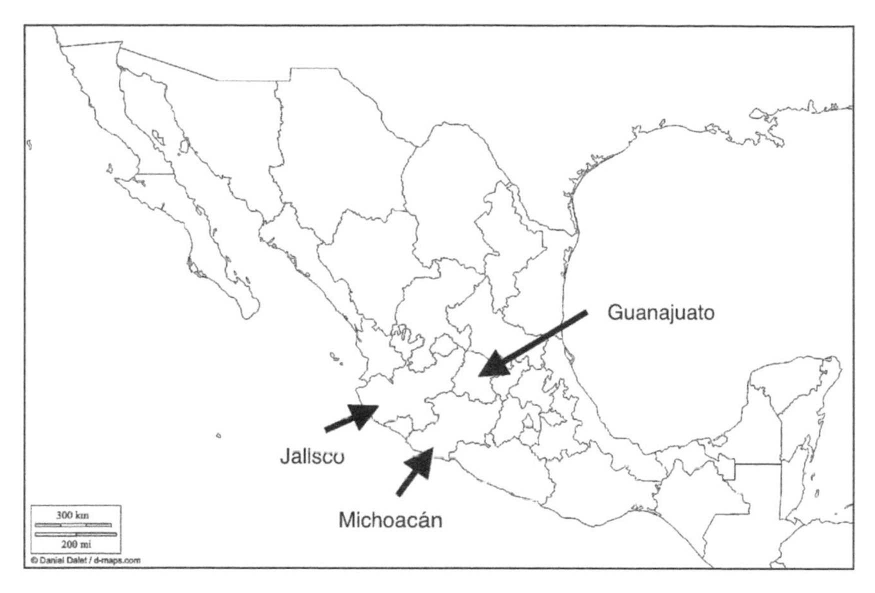

Fig. 1.4. Outline map of Mexico highlighting the three states sending the majority of Mexican immigrants to the Chicago area.

from the Distrito Federal, the district that encompasses Mexico City, and nineteen other states.[20]

Although conditions in Mexico caused many to migrate, the U.S. Immigration Acts of 1921 and 1924 were important pull factors in the continuing demand of Mexican workers in post–World War I South Chicago industry. Concerned that the end of World War I would lead to a surge of "undesirable" immigrants from eastern and southern Europe, Congress passed the Immigration Act of 1921. The law restricted immigration in any given year to 3 percent of the total number of foreign-born immigrants of each nationality residing in the United States as determined by the 1910 federal census. Because of the large number of foreign-born Mexicans already in the United States by 1910, the 1921 law allowed for the legal entry of a large number of Mexicans.

The Immigration Act of 1924, also known as the Johnson-Reed Act of 1924, reduced the quota from 3 to 2 percent of any nationality as recorded in the 1890 federal census. This change was directed at limiting eastern and southern European immigration while benefiting the more favored immigrants from northern and western Europe. In *Impossible Subjects*, historian Mai Ngai argues that the 1924 law demonstrated a shift from a cultural understanding of nationality to one determined by race. With the 1924 law, white Protestants descended from northern Europeans attempted to maintain their control of mainstream American society and culture.[21] Residents of Canada, the Canal Zone, and independent American countries were "non-quota" immigrants exempted from the limits for diplomatic and economic reasons. The exemption of Mexicans in the quota law allowed for the continued legal influx of agricultural, railroad, and industrial workers that had started over a decade earlier.[22]

In *Whiteness of a Different Color*, historian Matthew Frye Jacobson argues that "race is absolutely central to the history of European immigration and settlement" in the United States. Whiteness opened up the United States to these immigrants, "not any kind of New World magnanimity." It was, as highlighted above, race and the racialization of ethnic whites that shaped the Johnson-Reed Act of 1924. The keys to who got in and who did not, according to Jacobson, were racial difference and assimilability.[23]

If immigration was about race, why did Mexicans not fall under the quota laws? The answer is clear. Agriculture, the railroads, and eventually northern industry needed cheap labor. Bosses considered Mexicans least "invasive," most cooperative, and disposable when compared to racialized European immigrants. For racist and practical reasons, many bosses hesitated to go too far "down the color-line" and hire African Americans when available. In arguing that we must be careful not to make simple comparisons between the African-American experience and the racialized white immigrant experience, Jacobson writes that "it is not just that various white immigrant groups' economic successes came at the expense of nonwhites, but that they owe their now stabilized and broadly recognized whiteness *itself* in part to these nonwhite groups." In other words, it was the introduction and racialization of other groups that allowed the more established—and originally racialized—European immigrant groups to become white in the eyes of the dominant "white" society.[24]

Historian Gabriela Arredondo established that Mexicans in Chicago were racialized in much the same way as the ethnic Europeans before them. Looking at "ordinary life as the nexus for the creation of racist practices," she points out that it is not only how others perceived Mexican immigrants, but how Mexican immigrants perceived themselves that was crucial. In so doing, Arredondo convincingly argues that Mexican contact with European immigrant groups and African Americans was "critical to the formation of Mexican ethno-racial understandings of themselves and their place in the ethno-racial orders of Chicago." Throughout *Steel Barrio,* I explore these "ethno-racial" understandings among those inside and around the Mexican community of South Chicago while exploring how these understandings effected the environment and everyday lives of Mexicans in South Chicago.[25]

Mexican immigrants remained "non-quota" because of the intense lobbying by agricultural interests in Texas, California, and the American Southwest. The exemptions of Mexicans to the 1924 law sparked a backlash of prejudice against Mexicans, leading some nativist factions to press for the inclusion of Mexicans in the restrictive quotas. These efforts failed because of the powerful agricultural industry, primarily in California and Texas, which depended on Mexican immigrant field workers.[26]

Mexican immigrants acted to escape economic, environmental, and political turmoil in order to improve their everyday lives. Those who migrated to South Chicago learned of available work in there largely from family members, by word of mouth, or through *enganchistas*. Although some migrated for political reasons, the vast majority did so in search of a better life for themselves or their families. South Chicago industrial plant managers recruited Mexican workers and created labor shortages in part because of their racialized hiring practices. This complex combination of factors—all related to race, revolution, and riots—worked in tandem to create what would eventually become a vibrant Mexican South Chicago.

2

Finding Work

Gilbert Martínez learned about *enganchistas*, or labor agents, early in life, when Martínez's family moved from El Paso to the Midwest in 1917. Later, he recalled the trek north: "The *enganchistas* came in Pullman cars and they used to feed us sardines with crackers." His extended family was split up as workers and their families were dropped off in Midwest railyards. Martínez's immediate family and one uncle were heading for Illinois, while one uncle was dropped off in Hutchinson, Kansas, and another in Kansas City. Other workers went to Fort Madison, Iowa.[1] Although Martínez was probably confusing his family's *enganchista* with the men tasked with transporting the new hires to their Midwestern destination, his point is straightforward. The treatment of the new Mexican hires was much like the treatment of livestock.

Luis Franco entered the United States in 1923 to do farm work. That same year, he found a job through an *enganchista* working on a railroad just north of Houston. He spent twenty-seven days "with a pick and ax laying tracks" before his temporary stint ended. After an even shorter stint working as a *traquero* in the east Texas town of Cleveland, Franco went to Fort Worth in search of work: "As soon as we got there, we learned of trains leaving for the steel mills." The idea of moving to a new environment and into a new industry excited Franco. He signed a contract with the *enganchista* and boarded a "very long train" that dropped off people and train cars in Joliet, at the Carnegie mill in South Chicago, and in Gary. His stop was the bustling and polluted steel mill neighborhood of South Chicago.[2]

Those who transported Mexicans to the Midwestern worksites focused on minimizing cost, thus providing the new hires only minimal amenities and a small amount of substandard food. Contractors split up extended families depending on the labor needs per destination, with many workers learning of their destination as they were asked to get off the train. Mexican families endured these hardships because of the promise of new and better work that they hoped would eventually lead to a better everyday life environment at their new destination or to financial security in a post-chaos Mexico. Although not all of the details of Lucio Franco and Gilbert Martínez's journeys to South Chicago resonate in the same way for everyone who made the journey, their stories do reflect several broad patterns related to the role of *enganchistas*: traveling in the company of many other immigrants, some of them friends and family; using an indirect route to South Chicago; and a trajectory from railroad work to laboring in the steel mills.

Although the revolutionary influences in Mexico and the restrictive U.S. immigration legislation of the 1920s played very important roles in pushing Mexican labor into the United States and limiting the available worker pool for South Chicago industry, *enganchistas*—some of whom were independent businessmen while others were company employees—were essential to bringing a large percentage of Mexicans to South Chicago between 1916 and the onset of the Great Depression. Hundreds of Mexicans who came directly to South Chicago came as *enganchados*, or individuals already contracted to work in the steel mills or in the railroad yards of the South Chicago area. These men signed up with *enganchistas* at the Texas-Mexican border or deep inside Mexico, where they were promised free (or advances for) transportation to the work site.

For Mexican workers, the concept of using an *enganchista* to find work in Mexico or the United States was not new. Agricultural and industrial employers in the United States and within Mexico had used labor agents to contract Mexican workers for the railroads, the agricultural fields, and the mines of the American Southwest.[3] *Enganchistas* were particularly important for hiring Mexicans to work somewhere like South Chicago during the early phases of Mexican migration. South Chicago was a distant place in a strange environment with few Mexicans to send word of the jobs to friends and family.[4] Mexicans who signed on with a labor agent typically agreed to work for a company, be it a farm, railroad, or steel mill, for a minimum amount of time in exchange for guaranteed work and transportation to the worksite. Contracts for seasonal railroad or agricultural jobs sometimes included return transportation.

Historian Gunther Peck compared *enganchistas* with labor agents used in other ethnic communities in what he called a system of coercive labor relations in the North American West between 1880 and 1930. In *Reinventing Free Labor,* Peck focused on three *padrones*: an Italian immigrant based in Montreal who sent Italian immigrants to the Canadian West to work on the railroads, a Greek immigrant who lived in Salt Lake City and recruited Greek immigrants for the mines and railroads throughout the American West, and Ramón González, a Mexican American *enganchista* from El Paso who hired

Mexican immigrants to work in the railroads of the Southwest and the sugar beet fields of the Upper Midwest. *Padrone*, the Italian term for labor agent, was a common term used outside of Spanish-speaking communities and the term Peck chooses to use to identify the labor agents regardless of ethnicity. According to Peck, although all three of these labor agents found long-term control of immigrants difficult, González had the greatest difficulty because he was unable to control Mexican worker mobility, a key to a *padrone's* success.[5] Labor agents experienced greater difficulty in their attempts to control the mobility of Mexican immigrants under contract as compared to Greeks, Italians, or other European immigrants because Mexicans were closer to their home country and were near Mexican immigrant and Mexican-American populations in the American Southwest. In *Walls and Mirrors*, Historian David Gutiérrez illustrates how recent Mexican immigrants experienced tension with the residents of longer established Mexican communities in the Southwest and California, but could still depend on the community for information about jobs and housing.[6]

Despite the fact that new European immigrants also desired to improve their working conditions in order to achieve their financial and quality of life goals, they could not depend on the same level of ethnic support as Mexicans who were "foreigners in their native land."[7] Ethnic Europeans in the American West also formed communities in lumber and mining camps, as well as in groups of homesteaders, who were able to help newer immigrants. However, Mexicans in those areas that had been part of Mexico before 1848 were able to draw on the resources of more deeply rooted communities.

Throughout the United States, a significant number of Mexicans contracted to work on railroad maintenance-of-way crews "jumped" their contracts for more lucrative jobs with greater stability, a step many were willing to risk because if their decisions did not pan out, they had nearby resources on which to draw. Mexican worker mobility and willingness to jump contracts for better work benefited the worker and South Chicago industry. Mexicans preferred the better pay, the year-around work, and the stability of employment provided by the South Chicago area steel mills. South Chicago industry benefited from those who left their railroad contracts. The result was a large multi-industry influx of Mexicans into the neighborhood throughout World War I era.[8]

Before *enganchados* jumped their railroad contracts for points north and before *enganchistas* started hiring Mexicans for jobs in the Midwestern steel mills, Mexicans worked the American railroads. Railroad companies hired Mexican immigrants in droves as the drop in European immigration that came with World War I created widespread labor shortages.

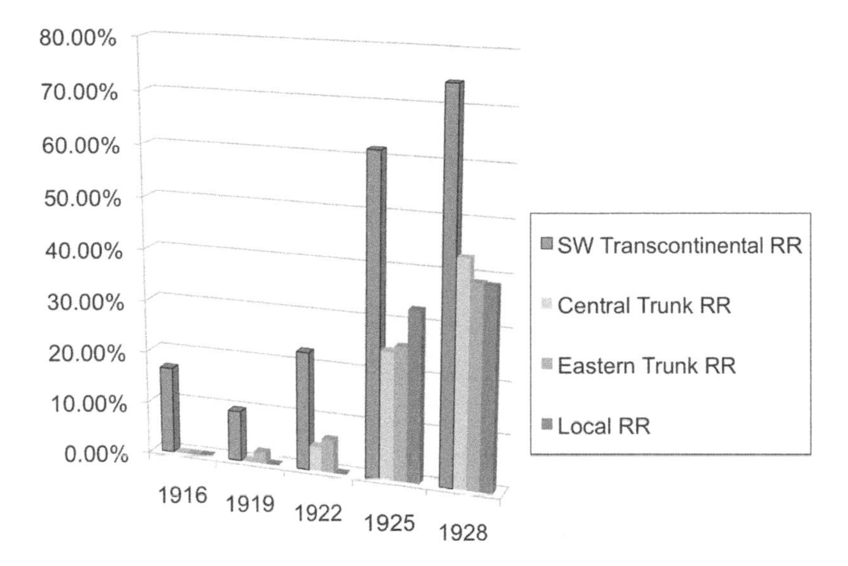

Fig. 2.1. Percentage of Mexicans in relationship to all employees on railroad maintenance-of-way crews, 1916–1928.

Even before the United States officially entered the war, American industry mobilized to provide materiel to allies in Europe. This exacerbated the labor shortage.[9] In 1916, railroad companies hired and transported 206 Mexicans to the Chicago area to work as *traqueros*. By the following year, 411 Mexican men were on railroad company payrolls in Chicago.[10] Although Mexicans were new to Chicago-area track work, railroad companies had used Mexican workers to build and maintain tracks in the United States for well over thirty years. The exact place of origin for many of these early Chicago-area *traqueros* is not known, but it is likely that railroad companies—companies that already had an established practice of hiring Mexicans—hired or transferred employees from areas like the American Southwest, Texas, and Kansas.[11]

When the United States entered World War I, demand for track workers increased as the movement of goods from other parts of the United States to eastern dockyards taxed the physical infrastructure of the railroad system. In December 1917, the nationwide labor shortage and increased demand on the railroads, combined with particularly bad weather and disorganized government military contracting, created a massive paralysis of railroad traffic in the Northeast. President Woodrow Wilson federalized the railroads later that month.[12]

It is not clear what happened to the early *traqueros* in the Chicago area after 1917 because surviving records do not identify specific individuals within this very mobile group. Moreover, they did not form an identifiable Mexican community within a city that by 1920 had 808,558 foreign-born immigrants in a total population of 2,701,705.[13] Even if all of these Mexican men stayed in Chicago and brought their families, they were scattered throughout the area in different railroad camps. This would have made area-wide or even cross-railroad-camp community organizing difficult if not impossible.

Although some Mexicans continued to work as *traqueros* throughout the interwar period, the steel strike of 1919 marked the starting point of a significant Mexican presence in the city of Chicago as a whole, and the neighborhood of South Chicago in particular. The strike, involving 365,000 steelworkers walking off their jobs in fifty-five cites and ten states, commenced on September 22, 1919.

U.S. Steel "imported" the first large groups of Mexicans to enter South Chicago during the strike. This strike was arguably the flashpoint that led to the influx of a large number of Mexicans in a short amount of time, but we should consider other factors to explain why U.S. Steel brought in Mexicans as replacement workers instead of other immigrant groups or African Americans. Steel mill managers turned to Mexican replacement workers to avoid having to hire African Americans in any significant numbers, a move some feared would provoke racial unrest similar to that of the recent race riots in other parts of Chicago.

The September 1919 strike was less than two months removed from what would become known as the famous Chicago Race Riot of 1919, a riot that left thirty-eight dead, over 500 wounded, and hundreds homeless.[14] Official coverage of the race riots in Chicago newspapers and coroners' reports portray the violence as exclusively black-white. However, Mexicans were involved in the race riots, including two men who brawled when white men attacked them, thinking the Mexicans were African American. Elizondo González was killed while José Blanco managed to knife one of his attackers, Joséph Schoff, who later died of blood poisoning. Blanco was "outraged" that he had been mistaken for an African American, an attitude that had as much to do with the racial politics of revolutionary Mexico as with the racial tensions among African Americans, immigrants from southeastern Europe, and Anglo Americans. Both González and Blanco's involvement in and omission from U.S. historical records of the riots capture Mexicans' ambiguous racial status in 1919—not white, and sometimes mistaken for black, in an elision that has largely been duplicated in scholarly accounts of the 1919 Chicago race riots.[15]

Not surprisingly, Mexicans are similarly absent from both contemporary and general scholarly accounts of the 1919 steel strikes. In his contemporary description of the strike, union leader William Z. Foster painted African Americans as having a pre-existing reputation of being "extremely resistant" to unions and unionization efforts. Despite acknowledging that unions shouldered blame for this attitude among African Americans because of the racism and discrimination within organized labor, Foster argued that African Americans "seemed to take a keen delight in stealing the white men's jobs and crushing their strike."[16] This thinking by the prominent labor leader illustrates the ingrained racialized antagonism that permeated much of organized labor of that time.[17] Scholarly studies more removed from the events and more attuned to some aspects of race relations in the United States discuss the ways in which African-American men viewed the strike as an opportunity to gain entry into an industry that had largely excluded them, as had the steelworker unions pushing the strike.[18] Standard analyses of the steel strikes recount that steel mills around the nation turned to more than 30,000 African Americans as strikebreakers and discuss African Americans' willingness to cross labor union picket lines.[19] Approximately 8,000 African Americans in Chicago used the strike as a gateway into industrial jobs that unions and business had long kept closed to them. These 8,000 men did not represent the extent of the available African-American labor pool, nor did it signify that African Americans filled the labor shortage. Steel mill managers remained wary of hiring African Americans in the wake of the race riots and used the riots as an excuse to continue excluding African Americans from skilled and semi skilled jobs in and around the mills.[20] This resistance to hire African Americans despite the labor shortage caused by the strike facilitated the recruitment of Mexican workers into South Chicago steel mills.

Despite the fact that the violence of 1919 did not spread directly into South Chicago, some steel mill managers were apprehensive about adding racial friction to an already tense labor situation. Some managers wanted to break the spirit of the strikers by hiring African Americans as replacement workers; others worried that such a tactic would cause rioting in and around the steel mills of South Chicago. Managers turned to Mexicans whom they perceived as less likely to spark rioting because less antagonism existed between whites and Mexicans than between whites and African Americans. As noted earlier, many Mexican workers were willing to leave the political and economic turmoil of a revolutionary Mexico for South Chicago. When Mexicans came to the area in 1919, many did not know that they had been hired as replacement workers. Others had their suspicions but were willing to enter the steel mills in search of better money and a better economic future. Although the steel mills hired some Mexican men from St. Louis and Kansas City, they recruited most of them from the Texas-Mexican border region.[21]

It is almost impossible to find a mention of Mexican strikebreakers in most scholarly treatments of the subject, and they are altogether absent from Foster's account of the strike. Foster possibly ignored them as their numbers were too low nationally for him to take them into account, or possibly he perceived Mexicans brought directly from Mexico as only temporarily in the United States and less of a long-term threat to the labor movement and to the striking workers.[22] Nonetheless, this absence of sustained attention to the role of Mexicans in the steel strikes of 1919 is indicative of a larger problem in most American labor histories: the reduction of discussions of race to the binary of white/African American. Only recently, as part of larger studies of Mexicans in the Midwest United States, have a few scholars investigated Mexican involvement in the steel strikes beyond a passing reference to them as strikebreakers.[23]

Steel mill managers chose their employees according to shifting racial hierarchies. Even as these hierarchies were based on mainstream ideologies of race, preferences for particular groups varied from one manager to another and also depended on the available labor pool. Managers who hired Mexicans did so out of a necessity created by their refusal to hire African Americans. A manager at East Chicago, Indiana, Inland Steel said he started employing Mexicans in significant numbers in 1919. At that time, Inland Steel sent employment agents, accompanied by Mexican employees, to Texas in search of workers. Advancing them their transportation costs, Inland transported around 150 single men plus some families in that first wave. They initially housed the Mexican *solos* in two or three bunkhouses while the few men with families settled in available housing near the steel mill.[24] A manager at Wisconsin Steel said he hired a few Mexicans for a short time in 1920, but complained that they went "south again at first touch of cold weather."[25] If in fact most of those employees did leave Wisconsin Steel, it was unlikely that cold weather was the determining factor. Although blaming the cold weather was a convenient and common excuse that deflected any blame from the foremen and managers, Mexicans in South Chicago did not tend to leave in the winter. In fact, the Mexican population of South Chicago usually grew in the winter as seasonal agricultural workers from throughout the upper Midwest regularly wintered in the neighborhood.

When Wisconsin Steel faced a shortage of workers in the spring of 1922, managers went to Kansas City after hearing of a surplus of Mexican workers there. They found a crowd of Mexicans looking for a job at the gates of Kansas City Nut and Bolt Company. Managers quickly started recruiting these potential workers. After being told by the superintendent not to take away their labor supply, one of those managers, Clyde Brading, hired a local Kansas City employment agency and put ads in the local Spanish-language paper. This tactic helped Wisconsin Steel hire 150 Mexicans from Kansas City. As with many contracted along the border, the new employees had to reimburse the company for travel expenses through payroll deductions.[26]

Brading commented that he found many Mexicans among the whites in Kansas City, but discounted this white/Mexican interaction by calling these whites "mostly the riff-raff."[27] Employers' preferences for Mexican workers over poor, white workers were not limited to the Midwest. Successful white professionals and businessmen "began to racialize poor whites as the 'scrubs and runts' of white civilization, both as an excuse to displace them and as a justification for the impoverished condition of those who remained."[28] In Texas, landowners and employers preferred Mexicans because they felt that Mexicans were more docile, worked harder, needed fewer provisions, and complained less than those that were part of the "white scourge."[29] As they did in Texas, many Chicago-area industrial employers turned to Mexicans because they believed that Mexican workers were cheaper, more dependable, and easier to deal with. They had become "the solution to the growing demand for a cheap and tractable labor force."[30]

However, not all managers in the Chicago area agreed on the desirability of Mexican workers. B. C. Mcleod, a manager at International Harvester's McCormick Works in Chicago, criticized Brading's use of and confidence in Mexican labor. He first employed Mexicans in 1925 to "dilute colored labor"—that is, African- American workers—and was unhappy with his experiences in dealing with Mexicans.[31] Richard J. Wuerst, a manager of the Chicago-area Illinois Malleable Iron Company expressed the most common negative stereotypes held by many employers who considered themselves forced to turn to Mexican labor against their better judgment to alleviate labor shortages caused by immigration restrictions. Wuerst, a native Chicagoan born to non-immigrant parents, described the Mexican immigrants who came to Chicago as a "downtrodden peon group" that was largely "scum" because "Hell, you can't do anything with them." Having employed "the Italians, the Slavs and Poles, then the colored" before bringing up Mexican immigrants from the Mexican interior, Wuerst complained of the high Mexican turnover rate. He also complained of the Mexican immigrants' desire for their own food even though the company had hired "an American cook for them, and put a third story on our building to make a rooming house for them."[32] These stereotypes and misconceptions reflected attitudes Mexicans encountered in South Chicago and further raised the tensions and discrimination they faced. The high-turnover rate and desire for Mexican food highlighted the mobility of Mexican workers and unwillingness to negotiate away a basic aspect of home culture: their food. Understanding *Mexicano* cultural values, their mores, and their transnationality is important to understanding the decisions *Mexicanos* made individually and as a community. Mexican immigrants, as well as Mexican Americans with close ties to Mexico, kept a foot in two worlds and appreciated the advantage of both. This priority placed on creating and recreating Mexican culture in Chicago facilitated a "third space" that helped shape Mexican South Chicago into a distinct Mexican enclave in the American Midwest.[33]

3

People and Patterns

Serafín García's mother and her family worked the sugar beet fields near Payne, Ohio, in the spring and summer of 1923, a job they had gotten through an *enganchista*. When the harvest ended, rather than returning to San Antonio, they went to South Chicago at the urging of friends. They had told García's mother to "'Stay over here" because "in the summertime you're closer to Michigan, you're closer to Ohio, you're closer to many places that you can go work on the farm." After three years of wintering in Chicago, Serafín García's mother settled permanently in South Chicago.[1]

Justino Corderos's migration to South Chicago also came through the encouragement of friends. In 1923, after working in a U.S. coal mine for about a year, Cordero went to South Chicago at the insistence of the Granado family who lived in the neighborhood. Cordero's *compadre*, Patricio Granado, took him to the employment offices of U.S. Steel. He got a job on a labor gang and ended up working at U.S. Steel for over forty-four years.[2] Cordero and Granado had both worked in Texas coal mines. They might have met there, but likely had mutual friends. At the time of the 1920 census, Granado was still working as a coal miner and living with his wife and kids in Texas.[3] By 1923, Granado was not only working at U.S. Steel in South Chicago, but also encouraging others to join him. By providing new immigrants with social and financial assistance through temporary lodging, knowledge about finding work and permanent housing, and orientation to life in and around South Chicago, people like the Granado family often remained the migrant's anchor, even after the newcomers had established themselves as Mexican South Chicagoans.[4] Over fifty-five years after Cordero's move to South Chicago, Cordero remembered Granado as his *compadre*. Serafín García and Justino Cordero's long journeys to South Chicago exemplify the importance of friends and family in encouraging migration and supporting the new migrants.

Events in Mexico and Chicago prompted an increasing number of Mexicans to enter South Chicago throughout the late 1910s and 1920s. They came

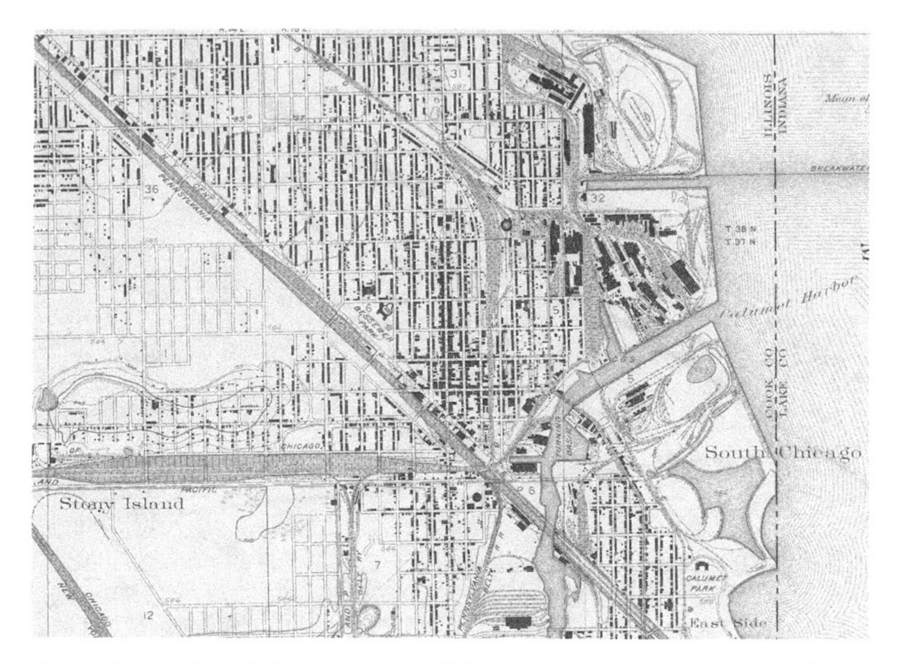

Fig. 3.1. Section of South Chicago closest to U.S. Steel, surveyed in 1927. The steel mill covers the entire lakefront. Note railroads clustering around the mill and through the center of neighborhood. Map courtesy of the University of Chicago Library's Map Collection, home of the original. *7.5 minute series (topographic): Chicago and vicinity*. Washington, DC: Geological Survey; Urbana, IL: Geological Survey Division, 1928–1929.

from Mexico and other parts of the United States. Many fieldworkers also came to South Chicago in the winter. Although individual and family motivations varied, Mexicans came to South Chicago for better-paying work, for community, and to join family who had preceded them. Yes, life in a polluted industrial zone was taxing, but living in South Chicago meant the possibility of economic prosperity and a comparatively comfortable living.

Even though the factors that prompted Mexican individuals and families to migrate to the Chicago area—and the routes they took—are almost as varied as the migrants themselves, discernible patterns emerge after a close examination of the migrants' motivations and routes. After early Mexican settlers to South Chicago created the initial groundwork of establishing physical and cultural Mexican communities, the migration networks grew exponentially. Chain and circular migration worked in tandem to create strong networks of new and returning Mexican migration to the area. The South Chicago Mexican community persisted through economic and environmental hardship. It grew in good economic times and contracted when there were few available jobs, but the community survived. Migrants continued entering the communities as long as they perceived that there were jobs available and the opportunity to make money, regardless of the actual availability of jobs.

Arguably the most popular Midwestern destination during much of the interwar period for Mexicans looking for work and a chance to prosper, the Chicago area provided employment, hope, and socioeconomic advancement for Mexicans, just as it had previously done for tens of thousands of European immigrants and African-American migrants from the American South.[5]

The role of *enganchistas* in recruiting Mexicans expanded as large employers continued to face worker shortages and Mexican workers' reputation as hard working, cheap, and loyal continued to grow. The steel mills of South Chicago used *enganchistas*, commonly in cooperation with their Mexican employees. Managers realized that the Mexican employees themselves were important recruiters who spread the word about available jobs to fellow immigrants throughout the Midwest, along the U.S. -Mexico border, and deep into the Mexican interior.

Despite distance, back-and-forth circular migration kept links between Mexico and South Chicago robust. Mexicans in South Chicago traveled to the border and into Mexico on a regular basis. This continuous back-and-forth migration between the Midwest and the Mexican borderlands became less common as the shift from seasonal farm work and seasonal railroad jobs to year-round industrial jobs made regular travel much more difficult.[6] Despite the fact that circular migration was not as common in Chicago as along the border, and diminished as Chicago-area Mexican communities matured, the familiarity and acceptability of this type of migration pattern reinforced a sojourner attitude and sense of mobility.[7]

Overall, migration networks and migration patterns are critical components in explaining who migrated and why those who decided to migrate chose South Chicago. Although *enganchistas* and employer-initiated hiring drives stimulated the steady Mexican immigration to the region during the late 1910s and 1920s, personal encouragement and support from Mexicans already in South Chicago to those back home to join them maintained and expanded the flow of migration. The majority of Mexican immigrants and Mexican Americans came to South Chicago at the urging of family and friends. This meant that those who participated in this chain migration influenced the timing, direction, and scale of the migration.[8]

Even as these chain migration networks became the most common pattern of migration to the Chicago area from Mexico, from Texas, or from the

Midwestern sugar beet fields, the earlier form of drawing Mexicans from border areas and the Mexican interior continued to account for a significant number of Mexican migrants to Chicago. Large South Chicago–area employers, primarily the railroad companies and steel mills, continued to use their own recruiters and to contract with *enganchistas*. This practice remained until the Great Depression.

Chain migration was not always in opposition to, or competition with, the *enganchista* system. The combination or cooperation of both systems explains how large numbers of Mexicans and Mexican Americans migrated to industrial Chicago within such a short period. The *enganchistas* helped to reorganize and stimulate chain migration at times of low migration and high demand for workers.[9] Mexican immigrants who wanted to migrate to South Chicago to join family or friends but lacked the financial resources to make the trip on their own used *enganchistas* to get them to their desired destination. However, the use of *enganchistas* meant the workers had to sign job contracts to work in the Chicago area in order to gain access the services offered by the recruiter. The two systems, though, did not always work in tandem. They clashed when border-based, independent Mexican and Mexican-American *enganchistas*, paid by commission from Midwestern companies, competed directly with the self-perpetuating chain migration. Immigrants who signed contracts merely to get the transportation to Chicago and jumped their contracts soon after arrival created problems between the *enganchistas* and the employer. Too many of these jumped contracts could cost the *enganchista* his recruiting contract with the mill or other employers.

As with Serafín García's family, a large percentage of early Mexican and Mexican-American migrants to Chicago came from the sugar beet fields of Michigan, Minnesota, and other Midwestern farm fields. Many farmworkers discovered Chicago when they were unable to afford the round-trip to the border. Farmworkers who decided to winter in Chicago could usually depend on transportation to and work in the sugar beet fields the following season. *Mexicanos* in Chicago often found work as *betabeleros* through employment agents or directly through the agricultural companies. *Enganchistas* with offices in downtown Chicago reached Mexican workers by word of mouth and through the distribution of flyers, while sugar companies advertised directly in local Spanish-language newspapers. One April 4, 1925, advertisement placed by the Michigan Sugar Company in the Chicago Spanish-language newspaper *Mexico* heralded the availability of work for *betabeleros*. The company offered men, both *solos* and those with families, $23 an acre plus transportation from Chicago to the farm.[10] These Chicago-area

Fig. 3.2. Mexican beet workers (*betabeleros*), near Fisher, Minnesota, October 1937. Photograph by Russell Lee, Farm Services Administration. Library of Congress Prints and Photographs Division, call number LC-USF33-011388-M3.

employment agencies, many of which were located in the Canal Street employment district, also contracted Mexican men who were unable to find work in the steel mills and packing houses to work the railroads.[11] Many of these seasonal workers—*betabeleros* and *traqueros*—eventually returned with their families to Chicago to winter and to find work at the better paying steel mills and slaughterhouses.[12]

Social workers and charity officials worried about the treatment of farmworkers migrating to the Chicago area as well as about the accuracy of workers' expectations. In 1928, the director of Chicago's United Charities lamented that *betabeleros* had not been provided transportation back to Chicago from the beet fields—a benefit afforded some farmworkers only a few years previously. The official, identified only as Mrs. Kembell by the interviewer, remarked that families ended the season with roughly $90 for the entire winter. "The company tells them there is work in the big city. The word goes round that there is work and good wages," complained Kembell. She described the typical incoming farmworker's family as entering Chicago with only a blanket roll. The only housing available to them was decrepit "two or three rooms for $8.00 per month" where "the landladies [left] the Mexicans a vermin infested bed and a stove."[13] This type of housing had historically been reserved for the newest and poorest immigrants. Although friends and family welcomed the farmworkers and seasonal track laborers into Chicago, the influx during the winter caused economic hardship for the Mexican community, where those who couldn't find work in the city had to wait until the March thaw to return to the fields or look for work as *traqueros*.[14] The vermin-infested housing without indoor sanitation and the neighborhood's pollution that made the residents' environment taxing did not keep an ever increasing number of new and returning Mexicans from South Chicago. The poor conditions for those who came unprepared, without someone to stay with, and without sufficient money to last winter, overburdened the veteran social service organizations and further crowded the parts of South Chicago reserved for those who could afford no other place. The dirty, vermin-infested beds and the dirty, crowded community might have made for an unhealthy environment, but it nonetheless meant hope and survival for many who were looking for something better.

It is true that the migration and settlement patterns created by Mexican immigrants to South Chicago lie in the political and economic transformations occurring both within the United States and Mexico, but a close examination of the factors that led to Mexicans settling in South Chicago must include the immediate context of family and community that shaped people's responses to the pressures of everyday life.[15] Examining the stories told by the individual migrants about their lives and journeys to South Chicago is useful in developing a better understanding of the circumstances, emotions, motivations, and hardships that affected the everyday lives of those who migrated to South Chicago. The personal stories that follow reflect the various and distinct experiences of those that came to Chicago during the late 1910s and the 1920s.

Luis García, Ernestine Barrato, and Mercedes Rios came to Chicago via the sugar beet fields of the Midwest. In Michoacán, Luis García was a skilled shoemaker with nine years of schooling. Hired by an *enganchista* in 1923 and "shipped" by the American Beet Sugar Company from El Paso to Minnesota, García initially only wintered in South Chicago and spent that time working at Wisconsin Steel, returning to his job as a *betabelero* each spring. García believed that the United States was a better place to live if one had steady work, but Mexico was a better place to live when you were out of work. According to García, the money and the work were better in the United Sates, but Mexico offered a temperate climate and the idea that neighbors, friends, and family looked out for each other.[16] By 1930, García was living with his family at 8949 Buffalo Avenue and working as a yard laborer in a steel mill.[17] Also from Michoacán, Ernestine Barrato (at that time Ernastina

Hernandez) and her parents, Manuel and Teresa Santos, traveled by train to Laredo, Texas, in 1925, when she was about ten years old. They lived in San Antonio for several years until her family contracted with an *enganchista* to work as *betabeleros* in Michigan. By 1930, Manuel was working in a South Chicago steel mill and his family was living near the mill on 8400 block of Green Bay Avenue.[18]

In contrast to Barrato and García, Mercedes Rios was a Mexican American, or *Tejana*, born in San Antonio, Texas.[19] After moving from San Antonio to the sugar beet fields outside of Matawan, Minnesota, a young Mercedes Rios and her siblings helped her parents in the fields. After Rios's father saved up to buy a new Ford, they spent a short time in Iowa before moving to South Chicago in 1923 or 1924 for what was to be a temporary winter job for her father at Interlake Steel. Rios's first impressions upon entering South Chicago were of a friendly host family, deep snow covering the buildings, alleys and gardens, and the streets being so quiet that it seemed as if "there was a funeral in each home."[20] Although Rios was accustomed to the northern winters by the time she moved to Chicago, she expected Chicago to be a bustling urban environment night and day. Her rural experiences had not prepared her for quiet and eerie South Chicago streets on a cold winter's evening.

As mentioned earlier, Serafín García's route to South Chicago was also indirect. Originally from Guanajuato, his mother, aunt, and grandmother brought a four-year-old García and his two brothers to Laredo, Texas. They had left Mexico for a safer Texas during the height of the revolution. His father remained in Mexico; having been drafted into Victoriano Huerta's army, he was later captured by, and forced to fight in, Pancho Villa's army. He was executed in 1916 after returning to his home state of Guanajuato that was, at that time, under the control of yet another revolutionary army, that of Venustiano Carranza. García recalls attending school and enduring extreme poverty in Laredo, Texas, in 1916 and 1917, along with many others who were fleeing revolutionary Mexico. It took eight years of moving around Texas picking crops and spending winters in San Antonio before García's mother contracted with an *enganchista* to work—as a family—picking sugar beets in Ohio. After that first winter in South Chicago, they returned to the fields the following spring. García entered public school in South Chicago during the winter and withdrew in the spring so that he could return to the sugar beet fields with his family.[21] After a third season in the fields, García's immediate family settled permanently in South Chicago. Serafín García's story is noteworthy because female-headed households were rare among Mexican immigrants to South Chicago. Most immigrants to the area were either *solos* or part of a family group that included two parents. García's mother was the sole wage earner during the first few winters in Chicago, working "a laundry job" to support her family. The family lived on one income until his oldest brother, Luis, secured a job in a steel mill in 1926.[22]

José S. Rodríguez came to Chicago via New Mexico and California in 1923. His first job in the United States was in the copper mines of New Mexico in 1918. From there he worked in the grape fields of California's Imperial Valley, where he boasted of making "good money in grapes." He then moved to San Francisco and later to Garden City, Kansas, to work in the sugar beet fields at the urging of an uncle. He came to Chicago in 1923 as a *solo* and secured a job with Illinois Steel when *solos* far outnumbered accompanied men among the Mexican steelworkers.[23] In his interview, Rodríguez argued that all Mexicans had preferred to return to Mexico to live once the situation got quieter and the revolution ended. Although he made more money by staying in the United States, he longed for Mexico because "the work was too hard in the United States." This sojourner attitude permeated the Mexican community. Mexicans in South Chicago kept alive hopes of returning to a safer Mexico someday in a better economic condition than when they and their families left.

Francisco Huerta was part of the first wave of Mexican immigrants to Chicago and founded one of the area's first Spanish-language newspapers. His earliest memories of his 1916 arrival in Chicago were of cold boxcars infested with "bed bugs," an experience not that uncommon for the *solo* Mexican migrant to Chicago. Huerta was only seventeen. He made his way to Chicago after deserting Pancho Villa's army near El Paso a year earlier. Following a one-day job laying concrete in El Paso, he secured work on the Santa Fe Railroad near Topeka, Kansas. He survived part of one winter in an unheated boxcar in the Topeka-area railroad camps before learning about available work in Chicago. He signed with an *enganchista* in Topeka to work in a Chicago area railroad yard and traveled as one of seventeen *solos* in a boxcar. Unhappy with the boxcar provided by the company as his housing in Chicago, Huerta boarded with a lively Mexican family of circus acrobats who had lived in Chicago for "many years." After saving money for less than a year, Huerta wired ticket money for his mother, Serafía, and sisters Esther and Trinidad to join him in Chicago.[24] He opened a boarding house for Mexican laborers; with the help of a hired cook, his mother and sisters performed the daily tasks of cleaning, cooking, and laundry. The family of acrobats who had originally boarded Huerta sent him any laborers in search of housing, as they preferred to board only actors. Although the idea of living in a house full of acrobats must have been intriguing, the acrobats' unwillingness to board laborers is a good example of the intra-community classism that was not uncommon. These referrals, though, do demonstrate the cooperation across occupational divisions and are an example of the kind of intra-community networks that allowed people to survive within the community. Although Huerta's later business success was unusual, his travel and entry into Chicago were representative of a broader spectrum of experiences.[25] Just as the specific experiences and routes of these individuals' journeys to Chicago differed, collectively they demonstrate the wide range of travel and early encounters of the thousands of women and men who shaped the Mexican community of South Chicago. Mexican respondents to surveys in South Chicago conducted during the mid-1920s and 1930s corroborate the wide extent of experiences that helped define the community.[26]

The processes that facilitated Mexican and Mexican-American migration to the Chicago area included easier travel from the Mexican interior to the American Midwest, the use of *enganchistas* along the border and within Mexico, and the strong chain migration to Chicago's Mexican neighborhoods once the core of each neighborhood was established. With the expansion of railroads in Mexico by the early twentieth century, one facet of the immigrants' journey had become easier. These railroads made long-distance travel much easier, faster, and more reliable. As railroads connected the Mexican interior with the Midwestern United States, Mexican immigrants traveled longer distances than they had just decades earlier.[27] Some Mexican immigrants settling in South Chicago came to the border knowing that Chicago was their final destination, but many *Mexicanos* who settled in South Chicago came after years of migrant or transient work in other parts of the United States.[28]

Why did *Mexicanos* go to South Chicago? Despite its somber, polluted industrial landscape, South Chicago stood for economic opportunity and hope. Visitors, social-service workers, scholars, advocates, and the residents themselves described South Chicago, as well as the other two neighborhoods that Mexicans settled in, as dark, polluted urban and industrial slums. For the individuals and families who migrated from Mexico, from the sugar beet fields of the Midwest, and from other regions of the United States, this community was nothing less than social and economic opportunity. Not everyone was driven primarily by dreams of economic prosperity; some came to join family and friends, and others came to escape political and economic chaos in revolutionary Mexico. Whatever their motives or specific hopes, all came expecting opportunity, security and change for the better.

Scholars have long argued that immigrant groups to urban areas were not able to function as independently as they might have wanted to and were not as independent as they believed themselves to be.[29] This gap between desire and reality was a stark slap-in-the-face for the early Mexican immigrants to Chicago. Yes, Mexicans were likely to jump labor contracts and move to new jobs and new locations in search of more money and a better lifestyle. This did not mean that the economy was accessible to them at every point. Mexican *solos* and family groups were at the mercy of their employers in isolated railroad camps and on sugar-beet farms. Mexicans who worked in South Chicago faced discrimination and limited opportunities in the city, within the steel mills, in housing, and within the labor movement. Despite year-round employment in the steel mills, Mexican steelworkers were limited to a few positions and continued to face ethnic harassment and discrimination. Steel mill managers and Mexican steelworkers interviewed in 1920 and in subsequent years emphasized the limited opportunities for Mexicans immigrants. The mills employed Mexicans in the lowest paying, most menial jobs, with only a few rising to the position of foreman. Mexican immigrants to South Chicago in the 1920s and 1930s quickly realized that economic prosperity was not as readily available to them as they had believed when deciding to travel to the area. Although many Mexicans continued to see employment opportunities in the Chicago area as a source of hope, many of their aspirations remained unfulfilled.[30] These Mexican migrants were adversely affected by the biases of the prospective employers, as they could only enter the industrial workplace where, when, and in the capacity allowed by mill managers.[31]

Events such as the Mexican Revolution, the Cristero Rebellion, the race riots and steel strikes of 1919, as well as the passing of restrictive immigration legislation, encouraged migration of large numbers of Mexicans to Chicago and the rest of the American Midwest. The experiences and emotions that ran through the minds of the Mexican immigrants as they entered South Chicago between 1919 and the Great Depression were as varied as the stories of the men, women, and children who ventured so far from their hometowns in the Southwest United States and in Mexico.[32] Although some reacted with disgust and dismay as they settled into the polluted industrial zone that was South Chicago, many were excited to reach a place where there was a promise of economic prosperity and a comfortable living. With few exceptions, the first wave of Mexican immigrants to South Chicago consisted of *solos*—married or single men traveling alone—who had signed contracts on or near the United States-Mexico border to work in South Chicago. Unlike Mexican immigrants to the American Southwest, the early Mexicans traveling to the Midwest entered an area with no pre-existing Mexican communities. Without an established Mexican community, early immigrants went without resources of support mechanisms, such as mutual aid societies, contacts for housing and employment, or cultural support.[33] Understanding this first wave of Mexican migration to South Chicago is important to understanding and appreciating how a vibrant and continuous flow of migrants developed. Each wave of migrants played its own important role in developing a community that persisted economically and culturally despite the economic disasters and turmoil of the Great Depression.

II

COMMUNITY

4

Home and Work

In 1919, Yugoslavian immigrant Helena Svalina was one of the first merchants to welcome and extend credit to Mexicans entering the gritty steel mill neighborhood of South Chicago. She did this despite the reaction of her children at the sight of the new immigrants. Her son Nick recalled that as a child he was initially scared of the new Mexican immigrants because of the popular portrayals he had seen on the big screen: "the pictures of cowboys and Indians, of Pancho Villa and all those guys in the movies." By 1919, Pancho Villa had come to represent Mexico and Mexicans in Chicago and other areas with few or no Mexican immigrants.

By the time Mexican immigrants had started entering South Chicago in any significant number, Pancho Villa had starred as himself in at least three U.S.-released films and documentaries, including movies D.W. Griffith filmed live on the battlefields of Northern Mexico. A showman enthusiastically supportive of the filming, Villa made a deal with a U.S. film company that included a 20 percent advance against royalties to be paid in gold. Many Chicagoans saw Villa "the bandit" as a representation of Mexicans and their culture. They had watched the brutalities of a real revolution on the big screen, and they had heard reports of Villa's "bandit" raids and subsequent run from General Pershing. Despite this, Villa was not solely responsible for Mexico's negative image. The popular 1913 film *A Trip Through Barbarous Mexico*, based on the book by the same name, is yet another example of a silver-screen feature that put Mexico and Mexicans in a not-so-flattering light. Offended by the movie posters and lobby cards for *Barbarous Mexico*, the Mexican consul to Chicago filed an official complaint with the city that the posters should be removed or covered. He charged that the posters, and the movie, denigrated Mexico.[1]

As *Mexicanos* continued to flow into the Chicago area, films like *Soldiers of Fortune* played in Chicago and were advertised in local papers. A 1920 ad in the African-American Chicago newspaper, *The Defender*, promoted

Soldiers of Fortune as an adventure movie that that featured "Mexican Bandits," "Revolutionary Chiefs," "Beautiful Maidens in Distress," and "Raids and Rescues." The top banner in the ad proclaimed, "Villa Bandits Enter the Movies." Despite the fact that Villa and his soldiers had been in the movies as early as 1913, movie promoters relied on the usual racialized tropes of Mexican bandits hurting maidens who had to be rescued by the white "fearless riders." Interestingly, promoters linked "outlaws" and "revolutionary leaders" to chiefs. Including these racialized and stereotypical Native-American terms further separated "them" from the white—and in this case African-American—"us."

Helena Svalina noticed Nick's reaction to her new Mexican customers and quickly assuaged his fears. She emphasized that they were nice people and there to stay and insisted that Nick and his sisters, Mary and Catherine, learn "Mexican." We don't know if Helena Svalina had her children learn Spanish out of a welcoming sense of humanity, out of a keen business sense, out of her understanding of the immigrants' journey, or out of some combination of those factors. We do know that Helena Svalina and her husband Sam were immigrants from Yugoslavia, entered the United States in

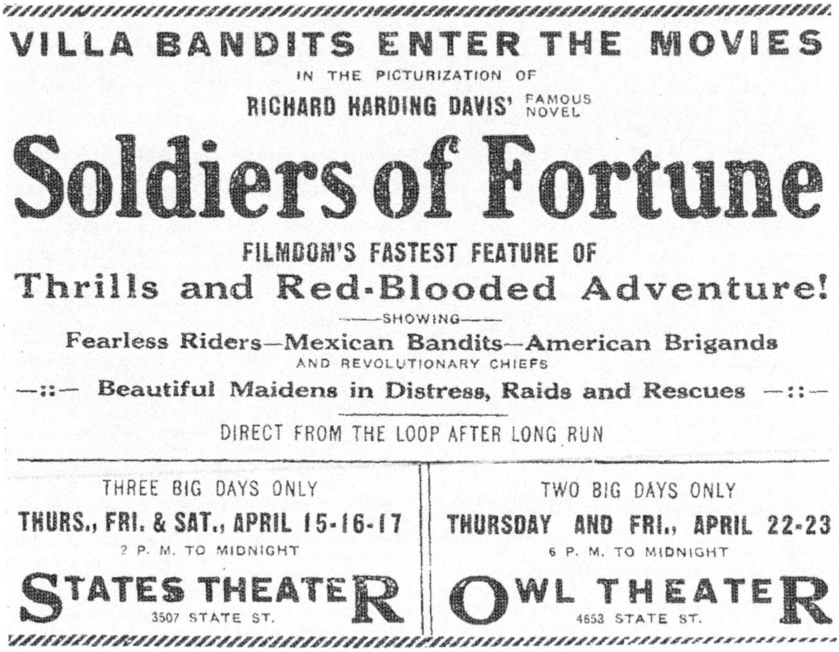

VILLA BANDITS ENTER THE MOVIES
IN THE PICTURIZATION OF
RICHARD HARDING DAVIS' FAMOUS NOVEL
Soldiers of Fortune
FILMDOM'S FASTEST FEATURE OF
Thrills and Red-Blooded Adventure!
—SHOWING—
Fearless Riders—Mexican Bandits—American Brigands
AND REVOLUTIONARY CHIEFS
—::— Beautiful Maidens in Distress, Raids and Rescues —::—
DIRECT FROM THE LOOP AFTER LONG RUN

THREE BIG DAYS ONLY
THURS., FRI. & SAT., APRIL 15-16-17
2 P. M. TO MIDNIGHT
STATES THEATER
3507 STATE ST.

TWO BIG DAYS ONLY
THURSDAY AND FRI., APRIL 22-23
6 P. M. TO MIDNIGHT
OWL THEATER
4653 STATE ST.

Fig. 4.1. Movie advertisement, *Chicago Defender*, April 3, 1920, p. 6.

1905 and 1907, respectively, and by 1918 already owned their home and a store with a butcher shop at 9001 Mackinaw Avenue. According to the 1930 census, Helena (listed in the census as Helen) and her husband, Sam, both spoke English and their native Croatian. In 1930, the census shows incorrectly that despite having been in the country for over two decades, they had not yet naturalized. Sam had in fact naturalized in 1921. It is intriguing that the 1940 census has both Sam and Helena with "first papers" but not naturalized. Historical evidence (a copy of his original naturalization card) confirms his status as a United States citizen; either the census-taker made a mistake or Sam and/or Helena did not disclose his status. Since Sam's status is incorrect in both 1930 and 1940, the most likely scenario is that they did not disclose his correct status to the census taker. It is clear, though, that the Svalinas achieved that ever-elusive American dream of economic and social stability that many of their new customers yearned for. In addition to being owners of one of the earliest shops interested in and frequented by members of the South Chicago Mexican community, the Svalinas eventually became landlords and bankers to local *Mexicanos*. Regardless of his immigration status or her motivations in helping *Mexicanos*, the Svalinas' outstretched hands aided newcomers' entry into the neighborhood while earning the Svalina family a prominent and comfortable living within what would soon become Mexican South Chicago.[2]

Nick, Mary, and Catherine Svalina were not alone in their fears of the incoming Mexican strangers into a South Chicago with a history of accepting western, eastern, and southern European immigrants. This was a working-class slum and a polluted steel mill neighborhood where residents, steel mill managers, and business owners maintained a perception of all darker-skinned Mexicans as ignorant, and at times lawless, peons. This was a neighborhood where immigrant and second-generation residents historically classified the newest immigrant group as less-than-white, less desirable, dirtier, and less intelligent then earlier immigrant groups.

By the time *Mexicanos* started migrating to South Chicago, the area had been home to Serbs, Lithuanians, Greeks, Swedes, Germans, Croats, Slovenes, Italians, Hungarians, Poles, Slovaks, Jewish immigrants, and the Irish. African Americans entered the area at about the same time as *Mexicanos* as the African-American Great Migrations occurred while the Mexican Great Migration was reaching the upper Midwest. Like *Mexicanos*, the vast majority of European immigrants entered the United States as less desirable, unskilled or low-skilled workers racialized as less-than-white. The steel mills and other industrial plants in the neighborhood made South Chicago a hub for "non-white" immigrant workers as early as the 1880s.[3]

The fact that the negative Mexican images and ideas were forefront in the minds of young and old South Chicagoans played an important role in the justification of Mexican discrimination. These stereotypes in the American Midwest had roots in the political rhetoric that came out of the United States war against Mexico, popularly referred to as the Mexican War. Democratic politicians racialized Mexicans in an effort to appeal to the Irish and to other European immigrant groups who had been excluded from mainstream Anglo-American society as "less than white." Democrats redefined the racial hierarchy of "white" and "colored" to move all Europeans into the "white" category. In this view of the world, Mexicans, members of a "mixed" race, were naturally and inevitably opposed to whites. This opposition made the war with Mexico natural, inevitable, and justified. Thus, Democratic rhetoric used the war to naturalize sentiment against Mexicans, a tactic welcomed by Irish Catholics as this new classification countered rising anti-Irish sentiment and included the Irish as "whites." In the 1858 Lincoln-Douglas presidential debates, Stephen A. Douglas appealed to Irish and other European immigrant voters in Illinois by contrasting Mexico's racial amalgamation to the "purity" of politics in the United States in which all participants were "of the Caucasian race," whether English, Scottish, Irish, German, or French.[4]

As illustrated by Nick Svalina's reaction to the new immigrants, South Chicagoans' prejudices came not from any personal interactions, but rather predated the mass migration of Mexicans to the area. Beyond film portrayals, these biases were part of a popular ethnic and racial ideology that classified each new immigrant group into a hierarchy according to factors that included country of origin, skin tone, religion, and perceived level of assimilation into the dominant "American" culture. In South Chicago, these prejudices manifested as harassment and discrimination in the workplace, in the availability and desirability of housing, and as institutionally imposed limits in social and recreational services. All of these manifestations—from the poorly ventilated and vermin-infested housing to the lack of access to recreational spaces and activities—directly affected the newcomers' health and the desire to quickly change their environment.

As *Mexicanos* in South Chicago worked to change their physical and cultural surroundings and to share resources and the use of physical spaces, they came to consider themselves part of a community that functioned to support its members, both newcomers and older residents, and used that sense of community to help each other overcome the environmental obstacles that reduced their quality of life. This sense of community also aided Mexicans in changing their surroundings to make life safer and more familiar.[5] As I examine these developments below, I also analyze the arenas in which

Mexicans formed community and connections: from within households or near them, when *solos* boarded with non-Mexicans, when Mexicans ate out to find familiar food and company, among fellow workers, and when family members chipped in to help the household survive.

Many factors worked in concert to create a physically and culturally distinctive Mexican environment in South Chicago. The physical location of Mexican settlement and the condition of housing made available to them by landlords and lodging-house keepers, the demographic composition of individual households, and the Mexican cultural mores brought to South Chicago by the immigrants are just a few. As one of distinct neighborhoods that made up the city, South Chicago was bordered to the east and southeast by

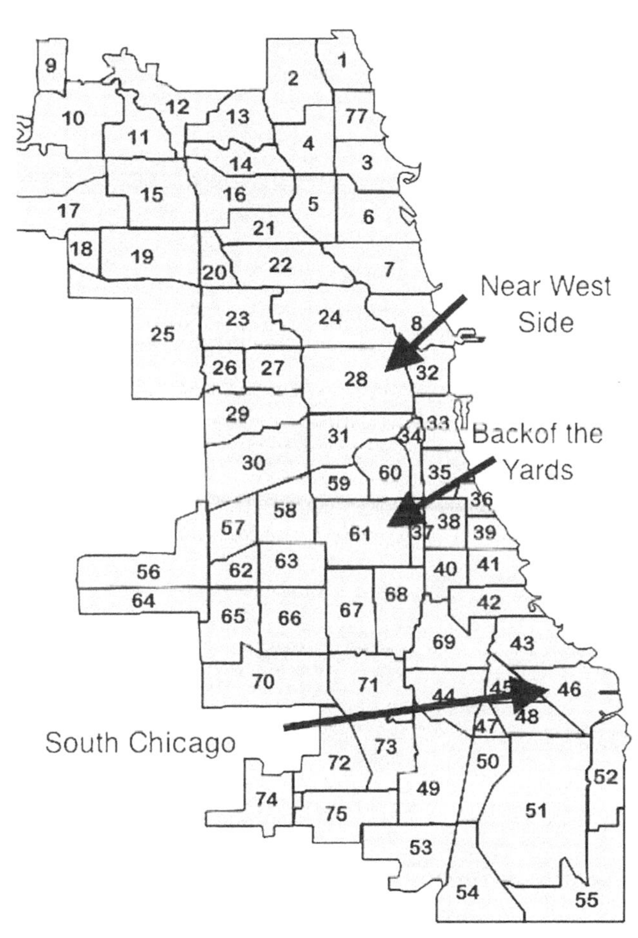

Fig. 4.2. Map of three primary Mexican communities in the City of Chicago before World War II.

Lake Michigan and Calumet River, to the north by 79th Street, and to the west and southwest by South Chicago Avenue.[6]

The South Chicago area became an industrial magnet and home for working-class immigrants starting in 1875, when work began on Chicago's first steel mill. It was opened in South Deering just along the Calumet River near 109th Street. Brown's Mill was a forerunner to Wisconsin Steel that was located in the same area. The first furnaces were fired up in 1881 at the location that would eventually become U.S. Steel's South Works. With the opening of Iroquois Steel across the Calumet River from South Works in 1890 followed by the predecessor to Republic Steel opening in 1901, the South Chicago area eventually became the locus of the largest concentration of steel-making facilities in the world.[7] The steel industry was always overwhelmingly dependent on foreign-born workers who were willing to take dirty, dangerous, and backbreaking work for low wages. The first foreign-born workers at South Works were mostly northwestern European (Welsh, Scotch, Swedish, and English). Eastern Europeans began migrating to South Chicago at the end of the nineteenth century.[8] By 1897, large numbers of Polish immigrants were working in the steel mills and living in the growing neighborhood surrounding the mills. Within a short fifteen years, they became the largest foreign-born group of workers in South Chicago. Polish immigrants remained the largest ethnic group in South Chicago throughout the 1920s and into the Great Depression.[9]

Massive buildings and towering smokestacks of modern American industry met Mexicans entering South Chicago for the first time in the late 1910s and 1920s. Steel mills and foundries dotted the landscape. For the single

Fig. 4.3. Ore docks, blast furnaces, and steel mills, South Chicago, IL, International Harvester Co., Chicago, IL, 1907, by Geo. R. Lawrence Co. Library of Congress Prints and Photographs Division, call number LOT 5786 no. 3.

men and families who came from the Mexican countryside, from the railroad yards of the Southwest, or from the sugar beet fields of the Midwest, the never-resting smokestacks spewing plumes of black smoke and ash framed by a grey, polluted sky, the large mill buildings, the stench, the rundown houses, the dirt alleyways littered with the refuse of everyday life, and the soot that covered everything were daunting and disorienting. The sight of smoke rising from smokestacks through the illumination provided by the furnaces' glare welcomed those who entered the neighborhood after dark. The newcomers' also experienced a feast of smells as their olfactory senses were not denied their own special welcome. The smell of slag, dumped by the mills onto the banks of Lake Michigan, added to the frequently overwhelming smells of a crowded, working-class neighborhood with poor sanitation and unpaved streets and alleyways. Immigrants were here despite the industrial pollution and poor living conditions to work in the steel mills or companies dependent on the steel mills. These mills not only dominated the local economy; they also physically and politically dominated the South Chicago landscape and waterfront.[10]

These South Chicago mills included Illinois Steel, Youngstown Sheet and Tube, Wisconsin Steel, Interstate Steel, and Federal Furnace. The economic fortunes of the steel mills in no small part dictated the economy of South Chicago and in turn the economic well-being of *Mexicanos* in the

Fig. 4.4. Postcard of steel mills at night, South Chicago, ca. 1915.

neighborhood. As primary employers in South Chicago, these mills dominated the environment in almost every conceivable way.[11] If *Mexicanos* were to survive in this foreign, urban environment, they would have to adapt and persist in the shadow of steel.

In 1911, only a few years before Mexican settlement, contemporary observers described a grim and grimy South Chicago that was a filthy, dangerous place to work and live. Sophonisba Breckenridge and Edith Abbott, leading Chicago academics and social workers, described South Chicago as an area with "great chimneys belching forth dense masses of smoke which hang over the neighborhood like clouds of darkness . . . so that no whiff of the air comes untainted." The neighborhood's air was tainted despite nearby Lake Michigan, the river, and the meadows surrounding South Chicago. So close to nature, yet so far from fresh air, the neighborhood was filled with "huge mills behind high paling fences" that blocked out most access to fresh air from the lake.[12] Changes to the neighborhood between the date of this observation and the early stages of Mexican settlement were for the worst as the population grew, industry prospered, and modern sanitation requirements went largely unenforced in South Chicago. The steel mills expanded and the neighborhood grew more crowded as demand for steel during World War I kept steel mills running at or beyond official capacity.

Breckenridge and Abbott described in fine detail the dismal conditions of wide "unpaved and unkempt" streets with a "dreary succession of small frame dwellings, dull in color, frequently dilapidated, uninviting and monotonous" on both sides. They painted a depressing and bleak picture of a soot-covered existence. An environment "made for industry, not for men and women and little children," overwhelmed visitors to South Chicago. Yet this was the daily reality for the immigrant working-class family.[13]

Breckenridge and Abbott went on to lament that with "magnificent enterprise" came a "hideous waste of human life" wherein "the men who feed the furnaces and send the products of their toilsome labor to a world market sleep in these miserable overcrowded houses" while having no decent places for relaxation and recreation aside from the "low saloons and dives" of South Chicago.[14] As social reformers and advocates, Breckenridge and Abbott voiced concerns over such a large concentration of young single men with no "appropriate" outlet outside of work.

Confining the newest racialized immigrant to the worst part of the neighborhood, in this case in the "Millgate" section, was not new. Working-class *Mexicanos* were simply the newest group that was given little choice by their bosses and neighbors but to occupy the worst part of the neighborhood. Although South Chicago was not the worst and most unsanitary neighborhood with a Mexican community—that honor was arguable held by the Back-of-the-Yards neighborhood made famous in Upton Sinclair's *The Jungle*—it is clear that part of this neighborhood was unkempt, polluted, and unhealthy.

Racializing the new *Mexicanos* and elevating most of the European immigrants into a "white" status made the same environmental racism perpetrated against African Americans acceptable against Mexicans in these industrial neighborhoods. By the time of the Great Mexican Migration to the northern United States, environmental racism had been alive and well in Chicago for decades. Historian Sylvia Hood Washington argues that "inferior whites or nonwhites . . . were prevented from occupying sustainable geographical spaces" and, when allowed to work, were confined to the most dangerous industrial jobs.[15] I argue that, even as most workers in the industrial neighborhood of South Chicago suffered from the environmental racism by city and industrial leaders, less-than-white *Mexicanos* suffered from additional environmental racism. The neighborhood itself was divided within the working-class community by level of whiteness, with *Mexicanos* and African Americans confined to the most derelict sectors of the neighborhood, withstanding the worst of this race-based discrimination. Dominant white society was responsible for and accepted industry's environmental degradation of these racialized neighborhoods, through the accepted negligence of property owners and government leaders, because they assumed that these nonwhite residents were by nature dirty and "used to" the poor living conditions.

Although less-than-white workers occupied many of the compromised neighborhoods, South Chicago and Back-of-the-Yards as examples, adjacent to their industrial places of employment, the status of the residents made continued environmental racism acceptable. As Sylvia Washington observes in describing Back-of-the-Yards: "Although there were not explicit legal policies or practices equivalent to race-based restrictive covenants or racial zoning laws that forced the immigrants into these spaces, the segregated geographical spaces that immigrants voluntarily created were disproportionately used as the final sink of the city of Chicago's waste as well as that of the packers."[16] South Chicago's environmental degradation and racism might not have reached the infamous heights of that in the Back-of-the-Yards; nonetheless the pollution and conditions in South Chicago would not have been acceptable in non-immigrant or non-African-American neighborhoods. This holds true when within the multi-ethnic/racial working-class neighborhood.

The physical conditions that awaited Mexican immigrants and Mexican Americans in South Chicago were bleak at best. Aside from the *solos* who slept in clean but generic bunkhouses on steel mill property or the railroad

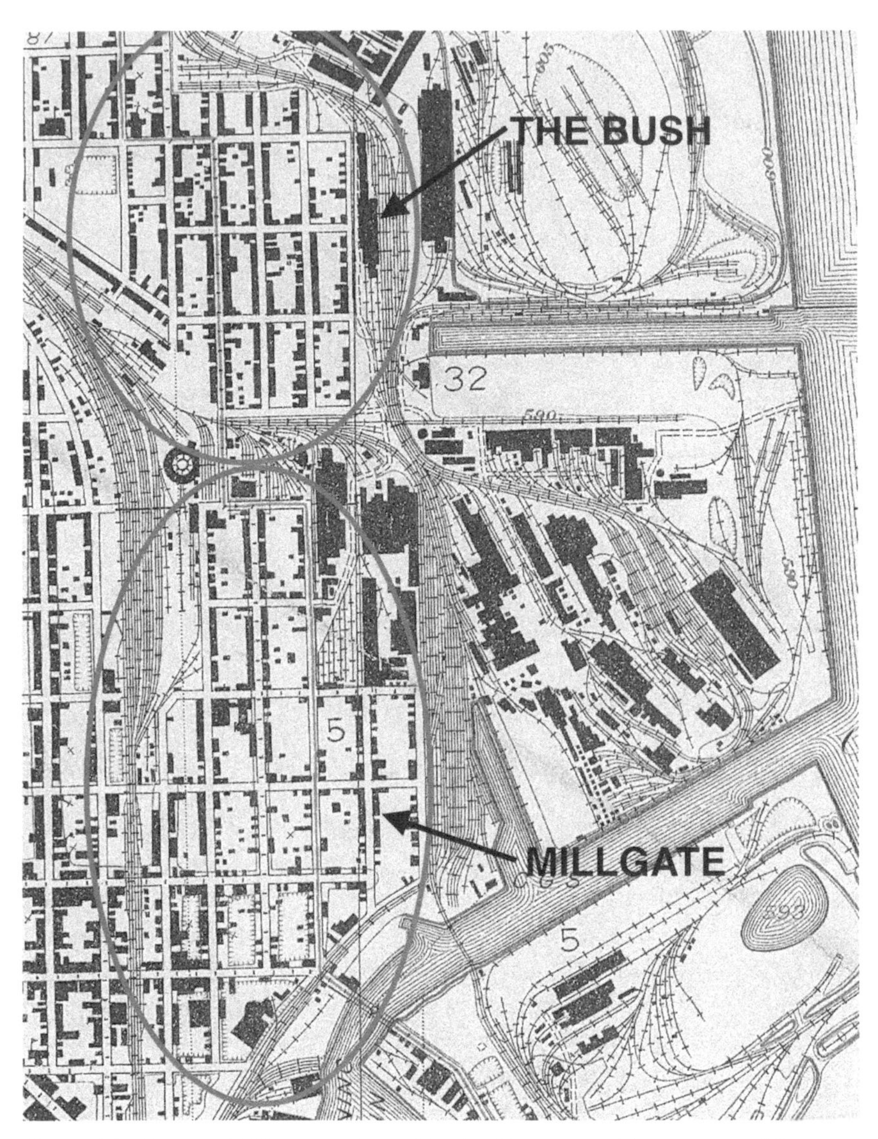

Fig. 4.5. The Bush and Millgate sections of South Chicago, 1927. Map courtesy of the University of Chicago Library's Map Collection, home of the original. 7.5 minute series (topographic): Chicago and vicinity. Washington, DC: Geological Survey; Urbana, IL: Geological Survey Division, 1928–1929.

workers and their families who slept in metal boxcars in camps that dotted the landscape, most Mexican immigrants had little choice but to settle in zones of transition near the industrial mills or other businesses that employed them. These zones of transition were physically as well as symbolically transitional for both the neighborhood and the Mexicans themselves. Mexicans not living on company property entered the only housing available to them, which in many cases were in areas surrounding factories and were in disrepair. In South Chicago, one of these areas was Millgate. Mexicans, who were in the midst of adapting to life in the South Chicago industrial environment were key actors in the transition of their immediate neighborhoods from entirely industrial to mixed industrial and residential areas.[17] Contemporary observer Elizabeth Hughes, a member of the city's Bureau of Social Surveys, found that by 1925, Mexicans and African Americans had "found shelter in the most used, most outworn and derelict housing which the city keeps." Mexicans lived in "The old tenement districts [that had] long been experiencing a steady encroachment by industry and commerce" and were residential sections "destined for extinction."[18]

Fig. 4.6. At 8921 South Green Bay Avenue, in Millgate, this was the last house standing between 89th and 91st Streets, and between Green Bay Avenue and Avenue O, when the photograph was taken, in 2006. The house was razed soon afterward. Photo by Curtis Myers.

For some Mexican railroad workers near South Chicago, home was a boxcar. Employers created neighborhoods, more accurately described as boxcar camps, to house Mexican workers and their families. These companies gave *solo* or accompanied Mexican railroad workers the option of living in boxcars. The 1925 Chicago and Western Indiana Railway (C. & W.I.) "boxcar colony" was located between 82nd Street and 83rd Street adjacent to railroad tracks. Anthropologist Robert Redfield commented that the neighborhood consisted of around twenty "very dilapidated box cars" scattered around repair sheds and railroad yard buildings. Many had iron-pipe chimneys, while some had porches, potted plants, and even small chicken yards. Workers and their families occupied thirteen of these boxcars.[19] Several other boxcar camps existed in the outskirts of South Chicago. A 1931 *Christian Science Monitor* article celebrated the large number of Mexicans in the Chicago area and described the sometimes long-term nature of the camps, claiming that

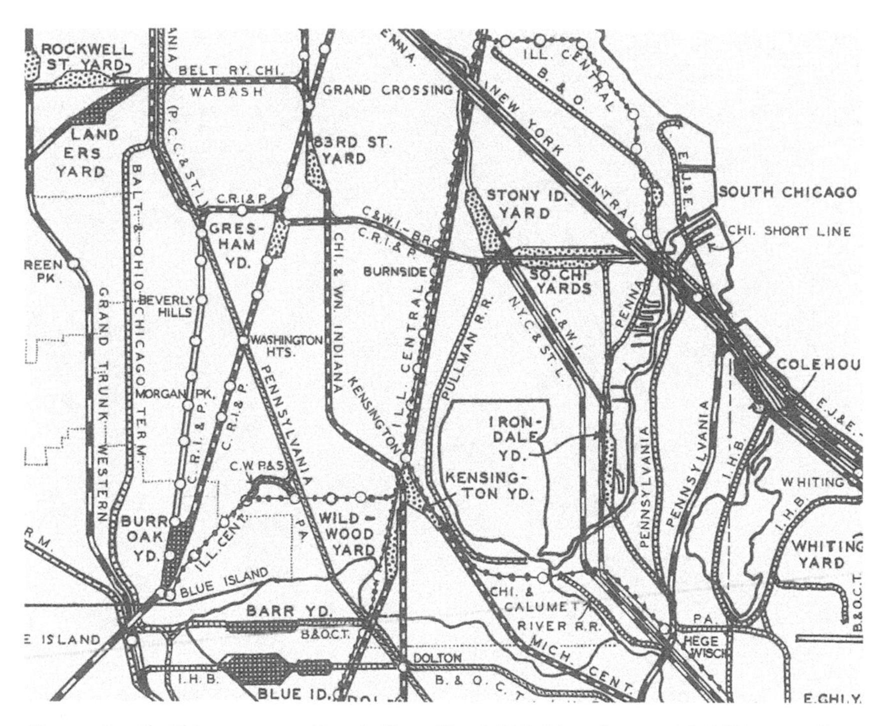

Fig. 4.7. South Chicago area railyards, from Harold M. Mayer's map titled "Pattern of Railway Facilities in Chicago and Vicinity," 1943.Courtesy of the Harold M. Mayer Estate. Original housed in the University of Chicago Library's Map Collection. Reproducing map in print or digital form is prohibited without the written permission of the Harold M. Mayer Estate.

some Mexicans stayed in their boxcars for "six, eight, and even ten years," because they were rent-free and provided homes that were "better than those in the crowded city" and "more like that to which the peasant were accustomed." Like Redfield, Kahn describes and praises the personalization of the boxcars, emphasizing that "some have gardens of flowers and vegetables" and that "the more prosperous of the workmen families have installed electrics lights, put up little porches and had tiny farmyards where chickens and perhaps a pig help out with the family's living."[20] Despite the implied temporary nature of the boxcar camps and the sojourner attitude of its residents, Mexican men and women living in these boxcars created a sense of community within the camps by improving the exteriors of their homes.

By adding porches, maintaining potted plants, and decorating the outside of their boxcars, the residents created a sense of pride, agency, and belonging despite their difficult economic and living conditions.[21] At first glance, the creation of a strong community seems difficult because of the temporary nature of the living structures, converted from the world of the railroad, their employer. However, these railroad workers and the women who accompanied them transformed those boxcars from a symbol of how the railroad companies dominated the lives of their workers, controlling their total environment, to an example of personalizing their environment and changing the industrial landscape to make it more human and humane. These alterations of their environment were one means of expressing connections among people, making visual allusions to a shared history and culture, and creating that shared culture in the face of larger differences.

Solos working for South Chicago steel mills during the early years of the settlement, from 1919 through the early 1920s, had the option of living in bunkhouses on steel mill property, as boarders in boarding houses, or with families in the immediate community.[22] When *solos* were given a choice about whether to live in the barracks-style bunkhouses or off company property, their preferences varied. Some preferred remaining in the mill-provided housing because of the abundance of prepared meals. A long-time Mexican resident of South Chicago interviewed in 1928 recalled that in the barracks "the rooms were very nice and the food we ate was good,"[23] though it was always American. He noted that "There was a variety and we never had too much of one thing but we never had tortillas, or menudo, tostadas, and the other things we liked." Some *solos* "made it a practice of staying out for supper and going to the families of some of the men in the neighborhood to eat" or would occasionally eat out for lunch because of their desire for Mexican food.[24] The quality of meals, though, varied greatly for *solos* renting rooms in the community.

This unnamed resident pointed out that another reason *solos* moved off company property was in order to separate themselves from the workplace during their off time. "After the day's work was done they wanted to get away from their work and forget about it and enjoy their leisure time. I remember one said to me once, 'this is too much like a prison. We work all day, then we eat in a big place all together, and sleep together. We never leave these walls of the factory.' I guess also we seldom had music or our theatre plays like the other Mexicans in Chicago did."[25] Many of the steelworkers who lived in the bunkhouses wanted to be part of the new and quickly expanding greater community of Mexicans. As more Mexican families moved into South Chicago, the number of *solos* moving out of the company bunk houses increased. By the early 1920s, steel mills started closing down their on-premises housing, forcing everyone to find lodging in the neighborhood.[26] During that same period, South Chicago Mexicans were starting musical groups, holding public dances, and staging plays at local schools and other venues. Many of the Mexican men in the bunkhouses wanted to be a part of this community and realized living outside of the mill gates would make that easier.

As early as 1920, Mexican single and married *solos* were living as lodgers and boarders in South Chicago. The 1920 manuscript census lists only a few Mexican families as South Chicago renters. Of the twelve addresses with Mexican lodgers or boarders immediately around the steel mill, Mexican families operated only two. Both, located at 9132 Burley Avenue, lodged Mexicans exclusively. The first lodging house, run by Eugene Navarro, a steel mill furnace-tender, and his wife, Felipa, had six Mexican and one *Tejano* lodger. All of them were steelworkers. The Navarros had two children—-one, a ten-year-old daughter and the other, twenty-three-year-old son who also worked at the steel mill. A thirty-three-year-old steelworker, his wife, sister-in-law, nephew, and two nieces occupied the other apartment at the same address, and four Mexican steelworkers were listed as lodgers.[27]

Mexicans also found housing with non-Mexicans. Those taking on Mexican lodgers and borders in South Chicago ranged from a Bulgarian saloonkeeper to an Austrian steelworker and his wife. In many cases, the Mexican boarders were *solo* steelworkers. In the zones of transition near the South Chicago steel mills, ethnic division took a back seat to the economic necessities of extra income for those taking on boarders and the need for inexpensive lodging for the unaccompanied male workers in the area. The occupations of the non-Mexican boarders in South Chicago boarding houses were as varied as their ethnicities and included, for example, an Italian bricklayer, a Michigan-born sailor, and a Greek steelworker.[28] One Bulgarian saloonkeeper in the 8900 block of The Strand had six boarders, one *Tejano* and five Mexicans. However, most boarding houses run by non-Mexicans had a much lower proportion of Mexicans among their renters. For instance, another Bulgarian saloonkeeper, this one at 9002 Green Bay Avenue, had nine boarders: another Bulgarian poolroom proprietor, a married Mexican steel mill worker, a Bulgarian barber, a married Greek steelworker, and three single Mexican steelworkers.[29]

Others running South Chicago boarding houses with Mexican boarders included: an Anglo-American woman living with two children and four boarders at 3219 East 92nd Street (a French-Canadian New Yorker who worked at a furniture store, a Michigan-born sailor, an Indiana-born worker in a car shop, and a Mexican steelworker); [30] and a German American couple with eight boarders at 9001 Brandon (four Mexican steelworkers, a sixty-year-old widowered Italian steel mill bricklayer, another Italian steel mill bricklayer, and an Austrian-American steelworker).[31] Mexicans lived in two other ethnic European–run boarding houses. One was at 9019 Commercial, where a German-American couple with a thirteen-year-old son lodged a German-American steelworker, an American railroad switchman, a Mexican steelworker, and a married Scottish steelworker with a nine-year-old son.[32] In addition, an Austrian steelworker and his Austrian wife boarded five steelworkers in their home at 8949 The Strand; they included one Mexican, three Serbians, and one Austrian.[33] These are all examples of boarding-house keepers in the least desirable parts of the neighborhood ignoring the prevalent anti-Mexican sentiment of the time to rent to Mexicans alongside other ethnic workers. Motivations and reasons had to have varied, but the fact that Mexicans were able to board in the homes speaks to the fact Mexican immigrants were allowed, if not welcome, in some sections.

As more Mexican families moved into the area, more of them took on boarders, further concentrating the *Mexicanos* of South Chicago. The most common living arrangement consisted of Mexican families living in two-floor apartments, reserving the first floor for the family and renting second-floor rooms to boarders.[34] As the Mexican population of South Chicago continued to grow, so did the density of the population, especially in the environmentally compromised area around Green Bay Avenue, Buffalo Avenue, and Avenue O, along the fence of the steel mill. In 1923, after only a few months of living in a bunkhouse, steelworker Alfredo De Avila moved off company property and boarded with a Mexican family on Avenue O. According to De Avila, the standard rate for a boarder that year was $18 every two weeks for the room and "lunch and dinner for sure."[35] Although many *solos* lived with Mexican families by this point, others did live in homes and boarding houses run by ethnic European immigrants and African Americans. In a 1925 study using a city-wide sample that included 266 Mexican households (as well as 668 African-American households and 590 white immigrant households), Elizabeth A. Hughes found that Mexicans had the highest proportion of one-family households containing lodgers. A full 43 percent of Mexican one-family households contained lodgers while 42 percent of African Americans, 28 percent of native whites, and 17 percent of foreign-born whites had lodgers. Hughes also noted that Mexican men were the most likely to be living in "non-family cooperative groups."[36] We can attribute the high number of lodgers in Mexican family homes to three primary factors: the fact that a large number of *solos* were entering South Chicago; the fact that many of these *solos* had come to the area with encouragement and support of more established members of the early-stage Mexican community in South

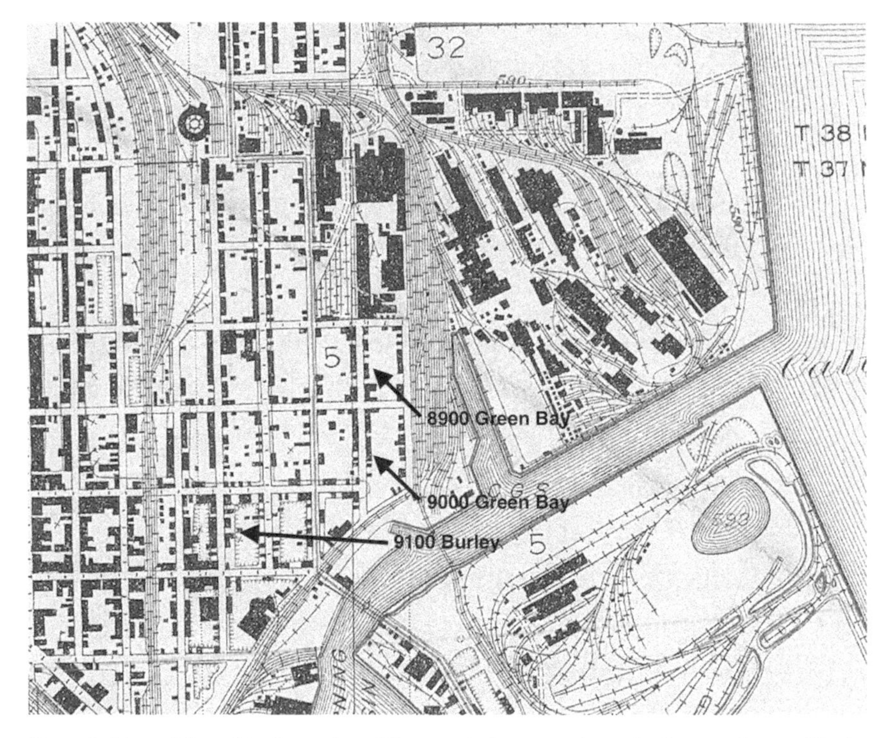

Fig. 4.8. Map of three blocks analyzed for population over time: the 8900 and 9000 blocks of Green Bay Avenue and the 9100 block of Burley Avenue. Map courtesy of the University of Chicago Library's Map Collection, home of the original. *7.5 minute series (topographic): Chicago and vicinity* Washington, DC: Geological Survey; Urbana, IL: Geological Survey Division, 1928–1929.

Chicago who then opened their homes; and the fact that rental income provided by boarders help support the host family.

The magnitude of the migration and the transition of sections in the neighborhood were dramatic. A scan of the 1920, 1930, and 1940 censuses shows streets that had a handful of *Mexicanos* in 1920 become dominated by *Mexicanos* by 1930. Including 1940 statistics in the comparison gives us a good idea of how a post–Great Depression neighborhood settled to numbers greater than 1920, but below 1930 levels. The 8900 and 9000 blocks of Green Bay Avenue and the 9100 block of Burley Avenue are representative sample blocks that show this change over time in the Millgate area of South Chicago.

The Mexican population compared to the total population on the 9100 block of South Burley Avenue was 21 of 185 in 1920 (11%), 96 of 271 in 1930 (35%), and 58 of 207 in 1940 (28%). Germans were the largest group on the block in 1920, Mexicans in 1930, and Mexicans slightly over U.S.-born whites in 1940.[37]

The 8900 and 9000 blocks of South Green Bay Avenue underwent dramatic shifts in ethnic makeup in this twenty-year period (1920–1940). Between 1920 and 1930, the 8900 block of South Green Bay Avenue went from 8 to 162 Mexican residents. They replaced Austrians, who all but completely evacuated the neighborhood, going from 155 in 1920 to 1 (a boarder) ten years later. The Yugoslavian population went from 0 to 92 for this same period. After the depression, the Mexican population dropped by 55% while the African-American population jumped by 57%. Austrians dominated this block in 1920, Mexicans in 1930, and African Americans in 1940.

As was the case on the 8900 block, Austrians controlled the 9000 block of South Green Bay Avenue in 1920 and disappeared by 1930. The Mexican population rose from 5 to 35 in this period, while the Yugoslavian population went from 0 to 57. The most dramatic shift was in the African-American population. 10 African Americans lived in this block in 1920; by 1930, there were 196. While Austrians were the vast majority of residents on this block in 1920, African Americans dominated the neighborhood in 1930 and 1940, even as their number dropped to 143 in 1940.[38]

What do these numbers indicate? The two blocks of South Green Bay Avenue provide interesting examples of the transition of part of a neighborhood "in transition." In this highly segregated city where African Americans and Mexicans were limited to the most environmentally compromised sectors, do these numbers indicate that this part of Green Bay was physically deteriorating or does it mean that the Austrians were assimilating and moving up the proverbial ladder? The most likely scenario involves a combination of both. African Americans enter the steel mills in greater numbers after the 1919 strikes, and Mexicans continue to flood into the neighborhood.

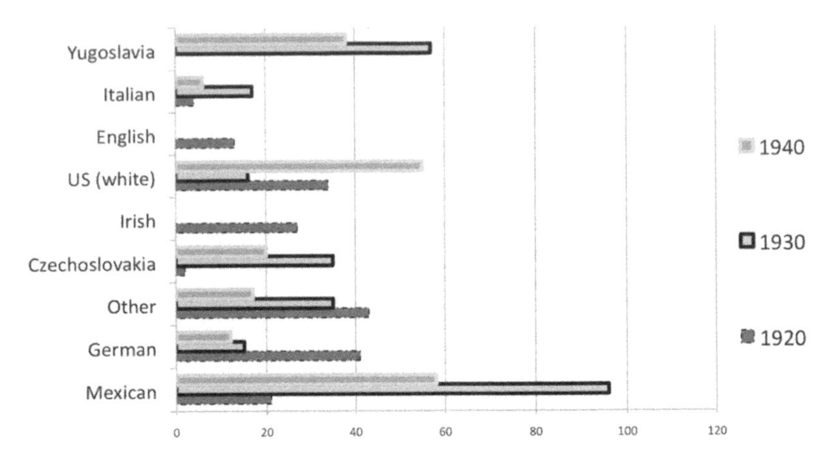

Fig. 4.9. Population of the 9100 block of Burley Avenue, 1920–1940.

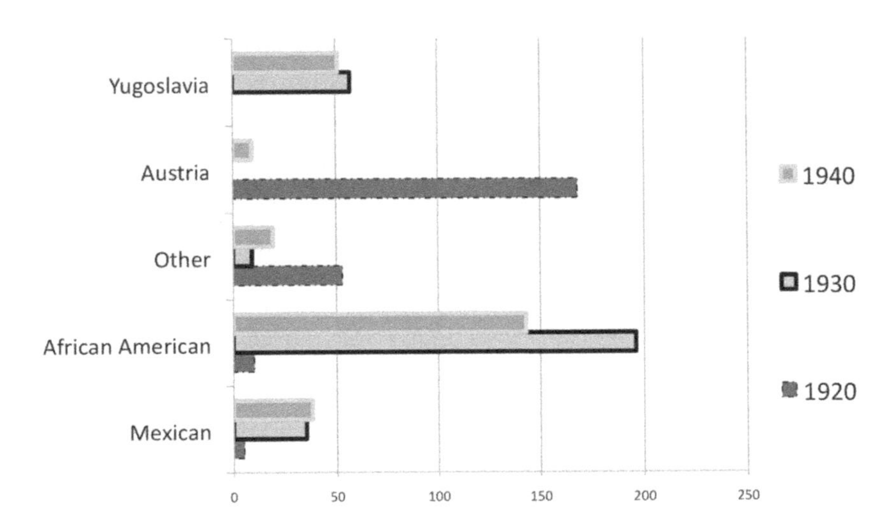

Fig. 4.10. Population of 9000 block of Green Bay Avenue, 1920–1940.

The house at 8921 South Green Bay Avenue serves as a great illustration of the transition of the 8900 Block of South Green Bay. In 1920, the house was split into two apartments that were occupied by Irish families. This was a time when European immigrants still dominated Millgate. An Irish family of four—an older woman and her three grown sons—lived in the first apartment. An Irish couple and their three young kids lived in the second apartment. All the men in the building worked at a steel mill.[39]

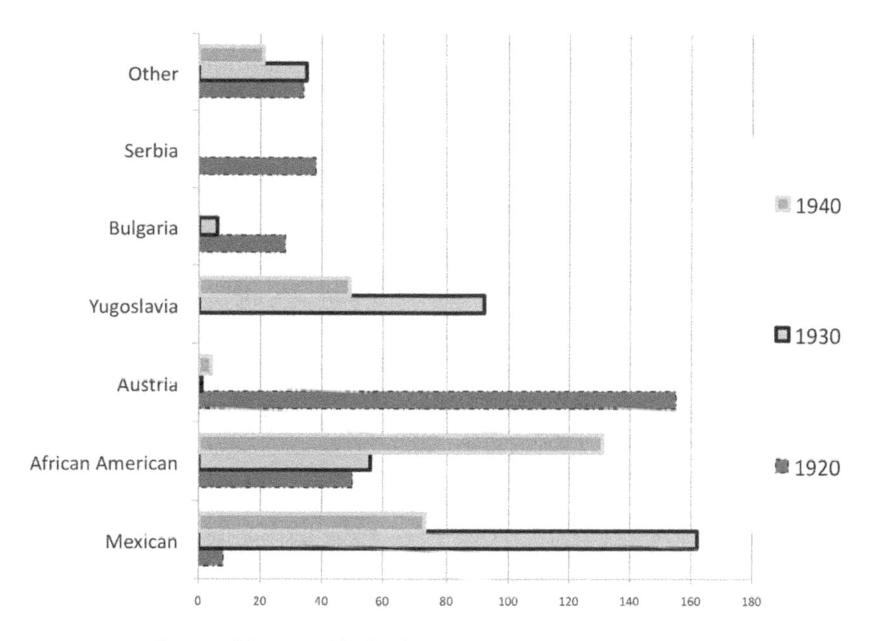

Fig. 4.11. Population of the 8900 block of Green Bay Avenue, 1920–1940.

Ten years later, as Mexicans entered the Millgate section in large numbers, only Mexicans resided in 8921 South Green Bay. The 1930 census listed the house as a single dwelling. This single dwelling, though, had twelve residents. Of these, ten were from the same family and two were boarders. The six working-age men in the house worked at the mill.[40]

As African Americans became the largest group on the 8900 block in 1940, they also took over the house at 8921 South Green Bay, which returned to its two-dwelling configuration. The first apartment had seven African-American migrants from Mississippi. The head-of-household, and only wage earner, was the fifty-three-year-old wife who worked "sewing." Also living in the apartment was her sixty-year-old husband, two daughters in their twenties, two granddaughters ages seven and nine, and a seventy-two-year-old woman listed as a lodger. Apartment number 2 also had seven African Americans—three from Mississippi, three from Tennessee, and one born in Illinois—and a woman who was head of household. She was thirty-three, with no employment listed, and had two teenage sons and a two-year-old granddaughter. Also listed was a thirty-three-year-old "roomer" and her teenage daughter. Nobody in that apartment was listed as having a job.[41]

Fig. 4.12. 8921 South Green Bay Avenue. The occupants of this house from 1920 to 1940 mirrored the transition of the entire block. Photo by Curtis Myers.

The 9100 block of Burley, on the other hand, was more acceptable to those on the white end of the racial spectrum. Despite being only a couple of short blocks from Green Bay Avenue, the 9100 block of South Burley Avenue had no African Americans enumerated in any of the three censuses examined. It is also important to note that this block was further away from the mill fence and on the eastern edge of Millgate. Although the Mexican community jumped significantly between 1920 and 1930, this block had many more U.S.-born Anglo Americans than the Green Bay samples. In fact, Anglo Americans were the only group to grow between 1930 and 1940. Also, this block had more Italians and Germans, two groups who represented more established—and "whiter"—populations in South Chicago.

While none of the recorded or transcribed interviews of South Chicago residents mentioned Mexicans living in boarding houses run by African Americans, the 1920 census lists at least two African-Americans who boarded Mexicans. Both African Americans were migrants from the American South. One, a steelworker from Alabama, lived with his wife, eleven-year-old son, and four Mexican boarders on the 9000 block of The Strand.[42] At 9007 Green Bay Avenue, an African-American steelworker from Kentucky boarded a

Croatian and a Mexican steelworker.[43] Additionally, two Mexican steelworker brothers rented an apartment at 8938 Green Bay in a building with two other apartments, both of which were occupied by African-American families.[44] The fact that Mexicans and African Americans lived in the same area is consistent with the reality that the most recent working-class migrant groups were forced into the worst available housing in the most polluted parts of the neighborhood along the mills' fences.[45] Despite evidence of Mexicans renting from African Americans, tensions between the communities did exist. Considered "less than white" but not black, Mexicans balked at being associated with the African-American community. David S. Weber argues that it was the presence of the Black migrants that allowed new European immigrants to be considered white, despite the pervasive nativism. It was this same presence, according to Weber, that allowed the dominant culture to accept Mexicans more quickly than was the case in the U.S. Southwest.[46]

The same marginalization and racialization of *Mexicanos* that made inter-racial interactions socially acceptable pushed *Mexicanos* into the worst housing and environment in South Chicago, making community and family networks an important part of everyday survival within the neighborhood and between South Chicago *Mexicanos* and those in surrounding communities. Not necessarily neighborhood based, these networks included family, were gendered, and varied depending on factors such as financial condition and level of assimilation. Important to the concept of these personal communities, or family networks, was their fluid nature. These extended household and family networks served as financial, moral, and cultural support during difficult times such as migration and unemployment, as well as a cultural community for personal celebrations and holidays. Maintaining status in these networks often required compliance to the generally accepted norms of the network, including those expected of the larger community. Thus, these networks reinforced Mexican cultural conformity through such channels as the use of Spanish within the community and a continued participation in Mexican cultural activities.[47]

In many cases economic and cultural survival required cooperation among families and households. The existence of cooperative strategies did not imply that a "master plan" existed at any level, but it did require constant negotiation between family members and others in the same household. This process of collective support was not exclusively monetary, but also consisted of goods and services. Collective support is admittedly a gendered concept, and therefore we need to consider the household support provided by non-wage-earning women in the family.[48] These contributions included household work—and household chores that became much more onerous when families took in boarders.[49]

Mexican households in South Chicago developed strategies that in time led to the development of a physical and cultural Mexican community in South Chicago. As a community, *Mexicanos* were able to change their physical environment by, among other things, opening and supporting Mexican businesses and businesses that catered to the community. These strategies and the development of the community reinforced the need for members to maintain cultural expectations. It was culture, and the pressure brought on by those "defenders" of the culture within the community, that defined which work or duties were typically men's and which were women's. However, gender transgressions, where men did women's work and women did men's work, were not uncommon. These changes in gendered roles, such as married women entering the industrial workplace or men having to do domestic work when boarding in a male-only household, resulted from economic or social necessity, as well as from the evolution of the locally distinct Mexican culture. These roles remained malleable as the community grew, matured, and changed.

One of many culturally reinforced expectations of community members was their continued support of family in Mexico. In addition to supporting children, the elderly, and other non-working members of the household, many South Chicago Mexicans regularly forwarded money to family back in their Mexican hometowns. Like most Mexican immigrants in the United States, Mexicans in South Chicago sent money to support family members in Mexico or to pay the sometimes dangerous and expensive passage for relatives who remained behind. These long-distance kinship relations were particularly important in that they distributed household risk by involving "mutual assistance" between those in Mexico and the Mexican immigrant households in South Chicago. Although life was financially difficult for most Mexicans in South Chicago, many continued to send money to family in Mexico via wire transfers.

Chicago-area bank officials' perceptions of Mexican bank customers ran the gamut from aggressively marketing to the Mexican community to not letting *Mexicanos* open accounts in their banks. An Atlas Exchange Bank officer in charge of Mexican community accounts, identified by Paul Taylor as Mr. Fernández, told Taylor that the bank, in the Near West Side neighborhood, had exactly 550 accounts owned by members of the Mexican community with accounts running from $5 to $600.[50] Mr. De Gerald, a vice-president of People's Stockyard State Bank in the Back-of-the-Yards neighborhood boasted of their enthusiastic courtship of the Mexican community. He claimed that his "business with the Mexican colony in this section of Chicago I dare say is as large or larger than any bank." He called the Mexican bank customer "very reliable and steady." Because of the racialization and marginalization of *Mexicanos*, De Gerald found it necessary to compare Mexican customers with African Americans:

> I would much sooner do business with a Mexican than a Negro any day. A Negro comes in, deposits a dollar or two and in a week they quit their job, leave the city and close out their accounts. They are the bane of our savings department. Not so the Mexicans. The Mexican when once they open an account and start to save, save regularly and well . . . We have over one thousand accounts with the Mexican colony in this section. . . . I believe we are fortunate in having the best of the accounts of the colony, for the Mexicans in this section of the town are here to work and live. They do not move around so much as I see some of the others do.[51]

In his comparison of African Americans and Mexicans, De Gerald expressed a socially prevailing stereotype of African Americans as lacking economic motivation and being shiftless and rootless. In fact, they moved for many of the same reasons that Mexicans did: in search of improved living conditions, because of discrimination, and in search of better jobs. By advertising in the Chicago's Spanish-language *Mexico* newspaper, de Gerald made it a point to aggressively "cultivate" the Mexican community living throughout the city and always had copies of *Mexico* at the bank to give to Mexican customers.[52] Regardless of the fact that Atlas Exchange Bank and People's Stockyard State Bank were not in South Chicago, they were representative of banks catering to the *Mexicano* community and sought customers from throughout the city.

Although the story of Mexican banking in South Chicago is less clear since Taylor and other contemporary investigators did not interview South Chicago bankers, a few other sources offer a sense of how important a welcoming bank was to Mexicans in the neighborhood. Mercedes Rios Radica recalled doing business at South Shore State Bank and interpreting for other Mexicans wanting to do business there. According to Rios Radica, the bank appeared to welcome Mexican customers, but employees did not go out of their way to accommodate Mexicans because they did not have any Spanish-speaking employees.[53] On the other hand, Union State Bank on 92nd Street and Houston Avenue in South Chicago did what People's Stockyard State Bank did. They advertised regularly in the *Mexico* newspaper welcoming Mexican customers. South Chicago Savings Bank advertised in the

Spanish-language *El Heraldo Juvenil* listing a specific cashier who could cater to the Spanish-speaking business community. [54]

As expected, payday was the busiest day for *Mexicano*-friendly banks. Long lines at the bank welcomed Chicago-area laborers on Saturdays. Both de Gerald and Fernández confirmed that Saturdays were their busiest days. According to De Gerald, in 1928, People's Stockyard Bank was selling 50 to 100 drafts a week ranging from "a dollar to a few hundred dollars."[55] Atlas State Bank, on the corner of Halstead Street and Taylor Street on the Near West Side, sold from 25 to 30 drafts on Saturday afternoon alone, with a total of 70 to 80 a week ranging from $2 to several hundred dollars.[56] Both bankers continually emphasized the Mexican worker's regularity in sending money to relatives back home.

Early Mexican immigrants to Chicago provided new and prospective migrants to the area with emotional and social support, but they also helped family and friends who had not yet made the move to Chicago. The fact that wire transfers to family in Mexico—placed through these banks or the post office—were common within Mexican communities of the Chicago area exhibit a rootedness to the Chicago area. This rootedness to Chicago, and the income earned while in the city, served as a means of connecting to and supporting family in Mexico, as well as facilitating the movement of family and friends to Chicago.

The evolving relationship between banks and *Mexicanos* was important beyond the business transactions between bank and customer. The willingness of several banks to cater to the community by hiring Spanish-speaking staff, by advertising in Spanish-language newspapers, and by providing those same newspapers for free in bank lobbies, speaks to the evolving importance of the Mexican community within each neighborhood. Although some banks racialized and excluded Mexicans, others near concentrations of Mexican workers viewed the welcoming of Mexicans as a wise business decision. For Mexicans, this commercial acceptance not only made sending money back home easier, but also provided recognition and status to members of the community. In addition, Saturday bank lines provided a safe space for Mexicans to gather and socialize while they waited their turn in line. Although workers might not have looked forward to waiting in long bank lines, these lines functioned as another space and place were Mexicans could gather without fear of harassment or discrimination.

Mexicanos entered South Chicago categorized as less-than-white, just like the Irish and Italians before them. Why were things different for Mexicans? Many Americans considered Mexico an uncivilized and barbarous place where bandits roamed and revolution ruled. Despite Helena Svalina's open welcome to her new customers and neighbors, most South Chicagoans racialized and confined Mexicans to the worst housing and the worst jobs. It was in this gritty, environmentally compromised neighborhood that the more established immigrant groups found power and a relatively elevated status as Mexicans came to occupy the lowest socio-economic stratum of the area. Living conditions evolved, not always for the better. By the late 1920s, Mexicans had become the majority in many Millgate blocks, physically the worst blocks in the neighborhood. Mexicans both adapted to their environment and adapted the environment to their needs. They opened businesses and started societies. From the seedy pool halls to the pharmacies and restaurants, street signs loudly proclaimed that Mexicans were here to stay. The long lines at *Mexicano*-friendly banks on payday provided social spaces for those on line as well as a statement of economic importance to those who passed by. While the Irish and Italians became "white" as they Americanized and became naturalized citizens, South Chicagoans continued to marginalize and racialize Mexicans, regardless of citizenship status, English language skills, or other markers of Americanization.

5

Great and Small

In 1929, seventeen-year-old Serafín García started working in the steel mills. Employed as a general laborer by Interstate Iron and Steel—later Republic Steel—García earned 39 cents per hour working ten hours a day, seven days a week. First assigned to a labor gang that cleaned out open hearths, he later moved to an outdoor labor position that provided steadier shifts and worked from 5:30 p.m. to 7:00 a.m. García took an outdoor labor position despite the terrible work conditions because it provided steady hours when compared to labor gangs whose hours varied with the mill's production cycle.

Alfredo De Avila's first steel-mill job was as a laborer building and moving the railroads tracks that were used to move slag from the foundry to the Lake Michigan shore on mill property. This was one of the least sought-after jobs at the mill and was frequently left for Mexican or African-American workers. The job required De Avila and others to endure both the brutal lakefront cold and the scorching summer heat while putting down and picking up railroad tracks as slag was distributed along the lakefront. Although working at the mill earned the worker credibility and respect within the Mexican community, members recognized this job as the bottom rung of the limited job opportunities at the mill. Not happy with this job, De Avila was able to find an indoor job after only a short time laying track. Once inside, he was assigned to a furnace room as a molder's helper. De Avila recalled that the company had two or three additional Mexican molders.[1]

Benigno Castillo started at Illinois Steel—later U.S. Steel—in April 1923 as a laborer moving dirt with a shovel. Castillo first entered the U.S. in 1919 at the age of twenty-one, having migrated from Techaluta, Jalisco, with his wife.[2] Although it is unclear when he migrated to Chicago, Castillo considered the $4.40 wage he earned for a twelve-hour day "a lot of money."[3] After only a week, Castillo moved to a contract position, paid by volume of production. Content working 8 a.m. to 8 p.m. during the week and 6 a.m. to 6 p.m. on Saturdays, Castillo stayed in this position until he retired.[4] García, De Avila, and Castillo are just a few examples of Mexican workers in the steel mills who had some—albeit, very limited—mobility within the mills. Some *Mexicanos* moved up to better positions, but managers promoted only a few Mexicans to foreman.[5] Just as railroad companies concentrated Mexican laborers in the lower-paying track maintenance jobs, steel mills limited the jobs available to Mexicans before the Great Depression to general outdoor laborer positions and to the most demanding furnace-related jobs.

It is doubtful that South Chicago steel mills would have recruited so many *Mexicanos* without the unintentional help of the railroad companies who introduced Mexican track workers—workers willing to jump their contracts for better jobs—to the Chicago area. By 1928, railroads grouped under "Southwest Transcontinental Railroads" had a Chicago-area track worker force that was 74 percent Mexican, while other railroads averaged only 39 percent. The railroads that had lines in the Southwest had the advantage of recruitment offices near large Mexican and Mexican American populations that funneled *traqueros* to Chicago. Although many Mexican workers left the railroads for better year-round work, the Mexican percentage of the total railroad workforce remained high because this industry served as a point of entry into work in the industrial Midwest, not least the steel mills of South Chicago.[6]

Once *traqueros* jumped their contracts for mill jobs, many *solos* and families traded their boxcar homes for the crowded and polluted neighborhoods with poor housing conditions in racially mixed blocks near the mill gates. Many Mexicans shared the same poor housing conditions as African Americans, yet many in the community hoped to maintain their distance for fear of being ostracized or likened to African Americans.[7] Despite the fact that most *Mexicanos* interviewed about their early experiences with African Americans in South Chicago said that both groups "got along," employers and neighbors who belonged to other ethnic groups frequently compared the "qualities" of Mexicans and African Americans—pitting both groups against each other for work and housing.

The arrival of Mexicans into South Chicago put them in direct competition with African Americans for unskilled steel mill positions. Mexicans, however, benefited from steel mill hiring practices that favored them over African Americans. Some steel mill managers justified discriminating against current and potential African-American steelworkers in favor of Mexicans because they worried that having Blacks in the workplace would put a strain on labor relations and would require the construction of separate eating and washing facilities to accommodate the de facto practices of segregation.[8]

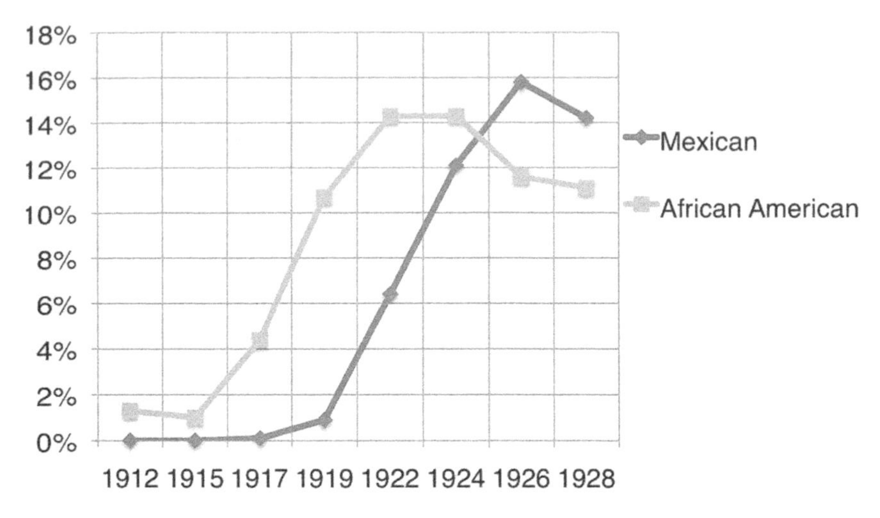

Fig. 5.1. Mexican and African-American steelworkers, 1912–1928.

Within a short time of Mexican entry into the steel mills of South Chicago, Mexicans outnumbered African Americans, not only because of the discrimination against Blacks, but also because managers created quotas to contain the growth of the African-American workforce.[9]

In East Chicago, Indiana, the Inland Steel Mexican workforce rose from just a few individuals in 1918 to 90 (1.8% of total workforce) the next year. Within two years, and after managers recruited and hired Mexicans as replacements for striking steelworkers, Inland Steel had 945 (18.6%) Mexicans on the payroll. The pre-depression peak, both numerically and by percentage of total workforce, came in 1926 when Inland employed 2,526 Mexicans (34.6%). Inland Steel consistently employed the most *Mexicanos* when compared to all the other plants in the Chicago-area steel and metal industry between 1919 and 1928. Industry-wide, Mexicans employed in the Chicago-area steel and metal industry jumped from 6 in 1915, to 18 in 1916, to a pre-depression high of 6,168 (14.1% of total workforce) in 1926.[10] Paul S. Taylor compiled ethnic and racial data from three steel mills (Inland, South Works, and Illinois Steel) to compare Mexican and African-American employment in the industry. His data showed that 266 African Americans (1.3%) worked in these mills in 1912. In 1919, 232 Mexicans (0.9%) and 2,699 African-Americans (10.7%) were listed. The following year, 1,228 Mexicans (4.6%) and 2,580 African Americans (9.8%) worked in these mills. Mexicans outnumbered African Americans in these steel mills for the first time in 1925, with 3,837 Mexicans (13.2%) and 3,717 African Americans (12.8%). As with the industry-wide sample above, the peak year for Mexicans was 1926, with 15.8%. African-American employment peaked in 1924 with 14.3% of the total payroll.[11] The percentage of African-American workers in the industry climbed slightly as Mexican workforce levels rose dramatically because Mexicans were the clear favorites to replace the dwindling supply of unskilled European immigrant laborers.

The workplace was full of the unexpected for many *Mexicanos* new to the area, to their job, or to both. Expectations and experiences differed greatly, even within the same line of work. Gilbert Martínez was hired by one of the area railroad companies as a water boy in his first attempt to get a job. Martínez recalled that the boss "saw I could speak English and all . . . even as skinny as I was I got a job as a water boy, just to talk to the guys. Bring the water and talk to the guys, that is all the boss wanted."[12] Work was plentiful, and Mexican crews did not hesitate to move from one line to another if they disliked their bosses. "One day the whole crew got mad at the superintendent, he was a mean guy. . . . He did not want to let them have lunch at noontime; some work had to be done. *Toda la raza* got mad at him. They left the job. [They] threw the shovels and everything. Just walked out."[13] Managers frequently complained that Mexican workers quit en masse when one of them felt they were mistreated or otherwise did not like the conditions. By "voting with their feet," Mexicans left intolerable situations, a tactic that gave them agency in an everyday environment that was fraught with harassment and discrimination by individuals and organizations at almost every turn.[14]

Although many Mexicans in the area worked for railroad companies, the steel mills were the primary employers in and around South Chicago and therefore directly shaped the community's environment and economy. This steel-driven economy naturally had a large part in defining how many Mexicans entered and left the community. The volume and availability of work at the mills set the overall tone for South Chicago, regardless of other industries or employers in the area. Although Mexicans are listed on steel mill employment roles as early as 1916, the first substantial migration, as emphasized earlier, was in 1919. By 1928, there were at least four Mexican foremen in the steel mills. One Mexican living in South Chicago described these foremen as old timers who knew English. He mentioned that the Mexican foremen were always given Mexicans to work with, thus segregating the Mexican foremen so that non-Mexicans would not have to take orders from a Mexican boss. Additionally, the unnamed resident recalled that "Some other old timers" had good jobs and did not have to "feed the furnaces or haul the bars."[15] These comments confirm the early consciousness within the community that limited advancement was possible, but learning English was a critical key in order to advance.

Some members of the community attempted to rationalize why Mexicans favored some jobs over others. "The Mexicans like steel because it's warm in winter. It is too warm in summer, but they keep the job so they will have it in winter." Working and being warm in the winter were important to many in the community who had first moved north for work in seasonal agricultural or railroad jobs. Others highlighted the fact that steel mill jobs were relatively stable and well-paying. Francisco Huerta, the Mexican newspaperman and local business owner introduced earlier, felt that Mexicans looked for jobs that were not as difficult, unless they paid very well. "They like the packing houses better than steel because there is not so much hard work. In the tin plate factory, there is no eating for 8 hours." He did say, though, that since "they pay piece work in [steel mill labor] gangs," workers could tell each other if one of them wanted to take a break.[16] This cooperation and solidarity among Mexican coworkers was not unusual within all-Mexican work gangs as they closed ranks to persevere in what were frequently hostile environments. The tone of Huerta's comments, though, bears witness to the heterogeneity within the community. His observations and consistent use of "they" were informed by his prominent middle-class status and a sense of separation between himself and the Mexican laborers he describes. Huerta was a newspaperman and boarding-house owner who had a sense of connection to these workers as part of the Mexican community, but this connection was undoubtedly paternalistic in nature.

While working in the steel mills during the 1920s provided more economic opportunities than other jobs available to Mexican men, these jobs came with the same workplace discrimination and mistreatment common in other industries. Mexicans regularly complained about the discrimination in the mill's employment office, the difficulties in advancing up the job ladder, and the lousy pay. Laborers of other ethnic groups—such as Polish, Irish, Croatian, and Hungarian workers—were slow to accept Mexican immigrants for reasons that included the inability of many Mexican immigrants to speak English, the growing level of ethnic prejudice, or the reputation of Mexican workers as strikebreakers. These attitudes were not limited to fellow laborers. The vast majority of foremen and managers, who were primarily Polish and Irish, refused to hire Mexicans to work in skilled or higher paying positions for many of the same objections voiced by laborers.[17]

Despite these impediments, *Mexicanos* in South Chicago endured and attempted to improve their everyday lives by finding ways to change their environment and form physical and cultural communities. In forming communities that were in part shaped by the demands on its members and in part shaped by the physical environment, Mexicans in the other areas of major Mexican concentration in the greater Chicago area—namely, Back-of-the-Yards and the Near West Side— as well as in East Chicago and Gary, Indiana—developed differently than Mexican South Chicago. As I touched upon earlier, many who settled in South Chicago lived near the gates of the steel mills. Mexicans who chose Back-of-the-Yards did so because of its proximity to work available at the slaughterhouses and the railroads. Meanwhile, the majority of those not working in the mills or the slaughterhouses chose to live in the city's Near West Side neighborhood because of its proximity to industry and downtown service jobs as well as the assistance and amenities available to them at Jane Addams's Hull House, the famous settlement house founded in 1889 that consisted of 13 buildings by the time of Mexican arrival. Housing choices within these neighborhoods were limited in the same ways they were in South Chicago. For the most part, Mexicans in Back-of-the-Yards and the Near West Side were pushed into the worst housing in the most unpleasant parts of the neighborhoods.

At the same time, the demands and external pressures on the Mexican communities varied. These pressures were shaped in part by the geographic distinctiveness of each community and by housing made available to *Mexicanos*. While some Mexican communities could expand as adjacent landlords made homes and lodging houses available to Mexicans, other Mexican communities, like the one in the "Bush" section of South Chicago, were limited by the natural borders of the neighborhood or by ethnic European neighbors who refused to rent to Mexicans. Work-related pressures—including the limited types of jobs given to Mexicans, the harassment and discrimination doled upon the workers by employers and coworkers, and the work schedules—varied from industry to industry and from one neighborhood to the next. The ethnic make-up of the immediate neighborhood, and the resulting inter-ethnic dynamics of that neighborhood, also made for significant variations in environmental and community pressures felt by Mexicans throughout the area.

Back-of-the-Yards, one of four neighborhoods in the Stockyard District, was settled mostly by Irish and German immigrants soon after the opening of the Union Stock Yard in December 1865, but had a significant Mexican population by the early 1920s. Upton Sinclair's 1906 novel, *The Jungle*, immortalized the area, also known as New City, as one of the country's worst slums. In part of a series of articles discussing working-class housing conditions in Chicago, Sophenisba Breckenridge and Edith Abbott compare South Chicago and the Stockyards District that included Back-of-the Yards less than a decade before Mexican settlement:

> South Chicago lies under the smoke-shadows of the steel mills, and in those mills are dangers to life and limb, blinding glare from the furnaces, magnificent exposure and terrible peril; but the influence upon the neighborhood is rather terrifying than degrading. In the Stockyards, on the other hand, are the mingled cries of the animals awaiting slaughter, the presence of uncared-for-waste, the sight of blood, the carcasses naked of flesh and skin, the suggestion of death and disintegration—all of which must react in a demoralizing way, not only upon the character of the people, but the conditions under which they live.[18]

This extremely dirty, degrading, and dangerous environment was home to generations of immigrants and some African Americans before Mexican entry into the area. Although officials implemented some sanitary improvements in the first two decades of the twentieth century, Back-of-the-Yards remained much the same as Mexicans took jobs in the stockyards and moved into the neighborhood. The city dumps, Packingtown, and the stockyards defined the northern boundary of the neighborhood located between 47th Street and 43rd Street. The Leavitt Street rail embankment and track, along with the city dumps, cut it off from neighborhoods to the west. The southern boundary was 55th Street, while the eastern end of the neighborhood ran along Racine Avenue to the stockyard.[19] The Back-of-the-Yards' population peaked at 92,659 in 1920.[20]

Despite antagonism and mutual discrimination between ethnic Europeans and Mexicans, social workers at the Back-of-the-Yards' University of Chicago Settlement House, commonly referred to as the Mary McDowell Settlement House, argued that the Mexican immigrants to the area did not compete with local, ethnic European immigrants because they took the least sought-after packinghouse and railroad jobs, the ones "which others do not want." To allay further Anglo-American and ethnic European fears about the new and increasing Mexican immigration, settlement house workers reasoned that Mexican workers were able to live on the low wages "by living in very congested quarters." These were the same houses and apartments that had long been reserved for the newest immigrants at the bottom of the neighborhood's pecking order. Like the railroad and steelworkers, many early Mexican immigrants to this packinghouse neighborhood came as *solos*. Some of them were contract-jumping *traqueros*. A settlement house report noted that "Those who [had] wives, supplement[ed] their salary by taking a number of their countrymen to room and board."[21] Experiencing comparable or even more severe social and economic restrictions as South Chicago Mexican families, Back-of-the-Yard Mexican families depended on the extra income from taking on boarders. Families depended on women's labor to run the boarding houses or the extra duties involved in taking on an individual boarder.

Like Mexicans in Back-of-the-Yards, those in the Near West Side neighborhood lived "wherever they [could] get in," often in homes "in the last stages of deterioration" with entryways facing alleys.[22] Like the other two neighborhoods with Mexican communities, the Near West Side was an entry point for unskilled and working-class immigrants to Chicago. The area surrounding Hull House was by far the most densely populated area in the city, and here, too, landlords and neighbors confined Mexicans to the most dilapidated sections of their respective ethnic, working-class neighborhoods, as they had the immigrant groups before them.[23] The Near West Side neighborhood included the area bounded by Kinzie Street on the north, the Chicago River on the east, Ogdon Avenue on the west, and the Burlington railroad tracks at 16th Street to the south. Although the Near West Side is best known as home to Jane Addams' Hull House, which was established in 1889, European immigrants settled in the area as early as the 1850s.[24] Just before the Great Chicago Fire of 1871, there were over 160,000 European immigrants living in the neighborhood, with the population rising temporarily to over 200,000 after the fire. Despite the fact that the Great Chicago Fire started in the Near West Side, it caused only minor damage to the neighborhood while spreading north and east. This facilitated its growth into the most densely populated neighborhood in the city.[25]

A University of Chicago Settlement House Report written in 1929 argued that congested living arrangements and poor conditions in Mexican communities actually benefited other groups. These poor and "unhealthful living conditions," the report argued, allowed ethnic Europeans to move up the social ladder by making the non-Mexican community "rather proud of the standard they themselves maintain[ed]."[26] Many members of European ethnic groups in Back-of-the-Yards and other Chicago-area neighborhoods believed that their own social standing improved when Mexicans arrived in Chicago.[27] At the same time, however, Mexicans experienced social and physical conflicts with their Polish, Irish, and Italian neighbors because the more established communities viewed Mexicans as a threat to their control of the area's physical and cultural environment. In South Chicago, these issues were evident in several parts of everyday life. Environmental racism allowed for bad housing conditions, as many Mexicans lived without hot water or indoor toilets despite the fact that outdoor privies had been outlawed before 1900. In the steel mills, Polish and Irish foreman forced and kept Mexicans in the most menial jobs. In South Chicago's Bessemer, Calumet, and Russell Square

Parks, neighborhood youth and park employees discouraged, and at times physically prohibited, Mexicans from using park recreational facilities and field-house showers.[28]

A few Northwest Indiana cities had sizeable Mexican populations. Like South Chicago, and unlike Chicago's Back-of-the-Yards and Near West Side neighborhoods, steel mills dominated these Indiana communities. Because of the common industry and the proximity of South Chicago and these cities, Mexicans from South Chicago frequently interacted with Mexicans from East Chicago and Gary. Located on the southern tip of Lake Michigan, East Chicago, Indiana, was only a few miles from South Chicago, just west of Gary. Mexicans first came to Gary and East Chicago at roughly the same time and rate as they did to South Chicago. Like South Chicago, East Chicago was dotted with boxcar camps that were homes to railroad workers, while being both economically and physically dominated by big steel. As with South Chicago, East Chicago's Mexican population first became appreciable after the 1919 steel strike. Inland Steel was the primary employer in East Chicago, and by the mid-1920s was the largest Mexican employer in the United States. In 1926, Mexicans comprised 34.6% of Inland Steel's total workforce.[29] Like South and East Chicago, big steel attracted Mexicans to Gary.[30] The large steel mills in Gary included Indiana Steel, National Tube, and U.S. Steel's Gary Works. Mexicans were 7% of Gary Works' workforce by 1926.[31]

By 1930, nearly 4,300 Mexicans lived in South Chicago, making them 7.6% of the neighborhood's 56,683 residents.[32] The city of Chicago as a whole, including South Chicago, had a Mexican population of 19,362, or .57% of the total population. Lake County, which encompassed both Gary and East Chicago, had 9,007 Mexican residents, making them 3.4% of Lake County's total population. Of those nine thousand, 3,486 Mexicans lived in Gary and were 3.5% of the population of 100,426; except for a few who lived in outlying rural areas, most of the rest—5,343—lived in East Chicago, comprising 9.8% of the population of 54,784.[33] While the method, timing, and rate of Mexicans' entry into Gary and East Chicago were similar to those of South Chicago, the percentage of Mexicans in the workforce in some plants, particularly in East Chicago, were much higher.

A key factor that linked Mexicans to each other, whether within a neighborhood or across communities, was the idea that they brought a common culture with them that could be used collectively to transform their environment. Arguably the most visible aspect of this culture was the use of Spanish both at home and in public. While speaking Spanish in public, placing Spanish-language signs and announcements on storefront windows and broadsides, and having the local "Mexican" paper available at the bank all served as lightning rods that attracted nativist attention and action, the Spanish language also served the critical function of supporting connections within the Mexican community of South Chicago and among area Mexicans. Spanish fostered a feeling of belonging among *Mexicanos*. Residents changed their physical environment with displays within the neighborhood—be it a loud public conversation or a poster on a storefront window—that helped demarcate a *Mexicano*-friendly space. Historian Mike Amezcua argues that post–World War II Mexican Chicagoans "all took meanings and made meaning of Chicago's streets and boulevards, places of leisure, places of work," and that it was "within those spaces, streets and buildings, hallways, and storefronts that [they] mediated their own cultural practices."[34] While agreeing in principle, I also argue that these dynamics were a legacy of pre–World War II Mexican Chicago. It was pre–World War II Mexicans in South Chicago and throughout the area who developed the expectations that *Mexicanos* needed to make meaning of and change their environment to persist and maybe even thrive in the ever-evolving community.

Mexicans had to find a balance in mediating their cultural practices. Some had a complicated understanding of what it meant to create an environment where Spanish was the dominant language. Too much Spanish use would impede economic and social success in the mainstream United States, where knowing English was considered essential. Although this premise assumes that an English-speaking *Mexicano* would not be racialized and would be accepted into the dominant white American society, the ability to speak English was key in moving into a better job with a higher level of financial stability. Language was not only a means to maintaining connections to Mexico and a common ethnic and racial identity, but also served as a marker of a sometimes unwanted difference between Mexicans and the dominant society. Despite the risk of a backlash, *Mexicanos* changed their physical environment by creating Spanish-language newspapers, by opening storefront businesses with Spanish names, and by celebrating Mexican patriotic holidays in public.

Many recent Mexican arrivals as well as long-time Mexican residents of South Chicago sought to maintain what they perceived to be the most important facets of their Mexican culture and tradition. This was South Chicago's "third space," the space that allowed for a Mexican South Chicago culture wherein what were perceived to be the best Mexican customs and the best Chicago influences could be selectively blended. Mexicans tailored their public and private physical environments accordingly. From the purchase of Mexican religious and cultural trinkets at a neighborhood store to the gendered cultural expectations shaping the behavior and role of women, men, girls, and boys, expectations and the community-imposed pressure to maintain a Mexican culture fashioned a distinct, but uniquely South Chicago Mexican environment. Some of these expectations came from community-wide pressures to conform and maintain Mexican cultural traditions. Mexicans' experiences in their homeland and in different Mexican communities in the United States shaped other expectations.

Learning and speaking English was a culturally charged issue with practical as well as symbolic ramifications. The fact that members of the dominant society demanded that Mexicans learn English, while many within Mexican South Chicago expected members to resist anything that might seem anti-Mexican—such as giving up Spanish in favor of English within the neighborhood—only further inflamed tensions. For many Mexicans, favoring English over Spanish within the neighborhood meant losing a little bit of the cultural unity that linked them with Mexico and with each other. For immigrant advocates as well as nativists, learning English meant assimilating Mexicans into the dominant culture and improving their ability to advance economically and socially.

As early as 1918, Presbyterian missionaries were active among Chicago's Mexican community with over thirty mission centers teaching English and other Americanization skills. "Teach Americanism in Box Car School" was one article highlighting this point in the *Chicago Daily News* in 1918. As Protestants fanned out in Chicago-area Mexican communities to "Americanize" the residents, converting some of the newcomers away from their Catholic roots was also part of their agenda. The Catholic Church did not move to establish a center to help Mexicans' religious and Americanization needs until March of 1921, when a center was built near a boxcar camp.[35]

Another contemporary English-language newspaper article provides an excellent window into the actions and attitudes of immigrant advocates and city officials working to improve Mexican life in Chicago through language instruction. An August 1928 *Chicago Daily News* headline proclaimed: "Mexicans Adopt Lincoln's Spirit: Bring Own Candles to School for Studies 4 Nights a Week." The reporter for this leading Chicago newspaper, who wrote of the hardships that Mexican immigrant laborers and their families endured, praised their perseverance and declared that "the spirit and method of Abraham Lincoln, who studied by candle light" was "not dead in Illinois." The actual dark classroom, though, loaned to the Chicago Board of Education, was nothing more than an old passenger coach in a railroad company's labor camp located in the outskirts of South Chicago. Adena Miller Rich, the long-time head of the progressive Immigrants' Protective League (IPL), invoked the spirit of Lincoln to explain that adult Mexican immigrants, with candles in hand, insisted that classes be conducted four nights a week instead of the scheduled two. Praising Mexican workers and their families as "eager for education," she commended the Board of Education for showing "an enlightened self-interest" by running these schools at a time when leaders in surrounding jurisdictions did not consistently provide English classes.[36] Though Mexicans in South Chicago guarded what they perceived as their indispensable cultural traditions that brought them together and linked them to Mexico, many did realize that learning English was important to get ahead socially and economically.

At first glance, the *Daily News* comparison of South Chicago Mexicans to Abraham Lincoln as a self-made man and patriot from Illinois illuminates the extraordinary energy and determination of Mexicans to become assimilated Americans and to succeed on American terms. However, the *Daily News* article also encapsulates the broad tensions in the Chicago area between the new and growing Mexican community and the Anglo-American and white ethnic residents in the years immediately before and during the Great Depression.

In this context, the Spanish language was a crucial factor in defining the Mexican community and staking out the physical and cultural Mexican environment. Although skin color did play an important role, non-Mexicans identified those in the Mexican community and considered them unassimilated because of the community's dominant use of Spanish. Many within the community considered the maintenance of the Spanish language a cultural necessity even as others identified English as a key to finding a better job and advancing economically. As external assimilation pressures increased, so did *Mexicano* defense of the use of Spanish. Residents continued to speak and display the Spanish language, turning the defense of the language into a form of resistance against the enemies of their community and culture.

The clear scene painted by the 1928 *Daily News* article—one of hard-working Mexican residents demanding and receiving instruction by the benevolent English teacher in order to become "American"—did not align with actual practices within Mexican South Chicago. In associating Mexican immigrants learning English with the ideology of Abraham Lincoln and the myth of his log-cabin origins, assimilationists were able to impose expectations while justifying the imposition of this language requirement. Simultaneously, assimilationists denied Mexican immigrants the resources and opportunities they needed to fulfill these expectations adequately. The article, which was published in a prominent, mainstream newspaper, reinforced the popular idea that immigrants were obedient, had to learn English, wanted to learn English, and could learn English at little expense to themselves or to the rest of society. In an Immigrant Protective League report produced in 1929, the author praised Mexican adults as "eager for educational opportunities" and thus "anxious" to attend night school.[37] Although imposing the legendary "spirit" of Lincoln into the immigrant sphere in Lincoln's own home state implied a drive to learn and the ability to overcome hardship, those imposing this idea did not directly associate Mexican immigrants with Abraham Lincoln, for that would imply the need to accept the Mexican community into the dominant American culture and power structure as equals. American society was not prepared or willing to accept the racialized *Mexicano* as one of their own.

By omitting the fact that both men and women attended, or needed to attend, English classes, the *Daily News* article highlights existing gender norms within the dominant culture and Mexican South Chicago. The comparison to Abraham Lincoln and the lack of reference to family implied that the universal "Mexican immigrant" was a man. The majority of men within both the Mexican community and the dominant culture expected women to remain in their homes serving their families and to stay within the traditional women's sphere of domestic housework. By making it difficult for women to learn English, men within and outside of the Mexican community were able to perpetuate the cycle that kept many women out of the classroom and in domestic activities. Mexican men in South Chicago complained that Mexican women who learned English were more likely to become Americanized and to lose their Mexican cultural identity. They also complained that women who worked or learned English and were exposed to the dominant American culture became "liberated" and thus would no longer be able to care properly for their husbands, children, and community.[38]

For some of the same reasons that Mexicans protested the assimilation of Mexican women and girls, assimilationists promoted programs specifically for them. Assimilationists targeted women and girls because of their roles as current or future caretakers and "keepers of the culture," while men were targeted to enhance their ability to gain employment and make social contacts that would improve their chances of becoming what advocates saw as productive members of U.S. society. In other words, assimilationist projects carried strongly gendered components. Adena Miller Rich, the Immigrant Protective League director and outspoken advocate for immigrant rights, argued for Americanization of all male and female immigrants to improve their chances of moving up the socio-economic ladder. Rich pointed out that the lack of affordable or free childcare for immigrant women made it difficult for them to attend English classes and that this in turn hindered their ability to get good jobs outside the home. Lack of basic education, even in Spanish, hindered Americanization programs. Some immigrant men and women came to Chicago from rural Mexican areas that lacked adequate schools and from families that kept girls from attending any sort of organized schools.[39]

Assimilation programs that targeted women were common in other areas with large Mexican immigrant populations. Historian George J. Sánchez points out that Mexican women in Los Angeles were targeted for early twentieth-century cultural assimilation programs. Just as American men had backed the education of nineteenth-century American women because of their role as mothers and caretakers of future citizen-sons, assimilationists used the ideals of republican motherhood to target Mexican immigrant women for integration into society because of their prominent role in the domestic sphere, including housework and raising children.[40] Those wanting to Americanize Mexicans believed that Mexican immigrant women had the greatest ability either to "advance [their] family into the modern, industrial order of the United States" or to "inhibit them from becoming productive American citizens."[41] Simultaneously, many male and female immigrants within the Mexican community believed that women had the primary responsibility for maintaining cultural traditions. Despite the fact that Sánchez's focus is on the Mexican community of Los Angeles, these Americanization and cultural concepts were not foreign to Chicago-area Mexicans and assimilationists.[42]

While gathering data on the number of Mexican women in the Chicago-area workplace, Paul Taylor uncovered stark examples of racialized bias and discrimination. One came from the Cracker Jack Company factory in Chicago, which, according its employment manager, had about fifty Mexican employees "of whom not over a couple are women." This manager favored Mexicans because they worked well in the heat and were "not so clannish as the Italians." At the same time, his "reason for not hiring Mexican women is that if they are very dark, they look dirty. The boss is particular, and so while they come around, we don't do much with them." Darker skinned Mexicans were associated with African Americans. In reply to Taylor's inquiry about any possible worker-based objection to Mexicans at Cracker Jack, the manager stated that "We ourselves don't want them" and added that "if colored apply, we simply tell them colored are not hired."[43] For Mexican women who did enter the wage-earning workplace, the most common jobs were in department stores, light industry, needle trade shops, and clerical work throughout the city.[44]

The preference of many in the community to limit the economic opportunities of women and older girls to traditional "women's work" complicated economic hardships by limiting household income. Even during times of economic hardship, many husbands and parents often discouraged Mexican women from finding work outside of the home, except within the specific spheres of culturally acceptable women's work. Although reformers and those in South Chicago's Mexican community believed that male vice and exploitation were to blame for the moral corruption of young women, these same groups were responsible for promoting an image of women as pure and passive, "demonstrat[ing] the vulnerability of young working-class women by denying their capacity for sexual agency and desire."[45] The types of generally accepted income-producing labor were centered on domestic work such as housecleaning and laundering. For most Mexican women who worked outside of the home, the added duties of wage-earning jobs did not spell relief from the amount of household work and other responsibilities to the family and the household.

In general, Mexican men in Chicago objected to their wives' entrance into the workplace outside of traditional, home-based "women's work." Although members of the community cited several reasons, the most frequently mentioned was the bad influence the workplace would have on women. Men feared women's "liberation," or an Americanization that provided more autonomy for women, and thus gave men less control over the daily lives of their wives.[46] Some women also disapproved of such autonomy for their daughters. Many more traditional Mexican men did not approve of their wives or daughters working outside of the home, or outside of jobs such as doing laundry, preparing food, or working in retail stores. Francisco Huerta used the distinctions in what was acceptable to define ethnic differences: "The Polish people put their women to work. The Mexicans don't want their women to work even if they are poor." Contrasting immigrant women with women born in the United States, Huerta said, "The Mexican immigrants don't like freedom of women, but it is all right for those Mexicans born here. The women find out that freedom of the U.S. is pretty good for them." His wife, identified only as Señora Huerta by Paul Taylor, disagreed: "I want to go to work, but my husband doesn't want me to." In Francisco Huerta's rebuttal to his wife's complaint, he spelled out the perceived fear of many Mexican immigrant men. It was okay, according to Huerta, to let American women, and Mexican women raised in the United States, to enter the wage-earning workplace, because they were "used to American customs." Arguing that his demand that his wife stay home did not stem from the jealousy she accused him of, he argued that "If our wives went to work, they would meet some other men and would go away with them; I would not blame my wife, I would blame only myself, because I have control of her. She would meet a man and go with him. Others are used to it, but not the Latin girls."[47] By putting his mother and sister to work in his boarding house, Huerta creates a false separation between acceptable domestic work and wage-earning workplace.

While cultural expectations within the community facilitated limitations on work that husbands and parents imposed on wives and daughters, men and the community as a whole also placed limits on the social behavior of women and girls. Many Mexican husbands and parents restricted the freedoms of wives and daughters in much the same way that nineteenth-century American reformers attempted to control the behavior of women who entered the workplace. They argued that Mexican women and girls needed protection from a more Americanized ethnic population that perpetuated such evils as promiscuity and led to the breakdown of the patriarchal household.[48] Despite these limitations, Mexican women did enter the workforce. Cognizant of their husbands' concerns and faced with the gendered and racialized limitations imposed upon them, many wage-earning married women worked in more culturally acceptable domestic-sphere jobs such as laundry, cooking, and housecleaning. Providing hardly any security and virtually no mobility, many of these jobs were considered temporary and used on a sporadic basis to contribute money to the household.[49]

The perceived need to protect young women and girls from the "evils of society" was not unique to Mexican immigrants. In *Delinquent Daughters*, historian Mary Odem argues that campaigns to protect the morals of young women in the United States started in the early years of urbanization and industrialization.[50] Although these earlier campaigns focused on the moral protection of middle-class white American young women who worked outside of the home, advocates set up missions in working-class neighborhoods in order to provide an evangelical Protestant-inspired ministry that focused on the moral protection of women and girls.[51] These campaigns reflected what was going on in Mexican South Chicago. Even as much of the morally based restrictions placed on Mexican women and girls came from within the community, South Chicago Protestant missions contributed through their neighborhood-based recreation and Americanization programs.[52]

The Catholic Church was also a presence in the neighborhood and a guardian of what many in the community considered "traditional morality." Although Mexicans attended area Catholic churches before 1923, no Spanish-speaking parish catering to the *Mexicano* community existed until then. In 1923, the archbishop of Chicago appointed the Spanish-speaking Jesuit Father William T. Kane to serve the Mexican community of South Chicago. Raising money to buy an old wooden-framed house at 9024 South

Mackinaw Avenue, in the Millgate section of the neighborhood, Kane converted the small building into Our Lady of Guadalupe Catholic Church, the first Chicago Catholic church created to serve the Mexican community.[53]

Subsequently, however, Father Kane was plagued with poor health, and so in 1924, responsibility for Our Lady of Guadalupe was handed over to the Claretian Missionaries, formally known as the Missionary Sons of the Immaculate Heart of Mary.[54] According to historian and archivist Malachy McCarthy, bringing the Claretians to Chicago "signaled the establishment of a more aggressive Mexican evangelization effort which would be closely allied with both the Catholic Instruction League and the local Irish community." Founded in Spain with missions in the American West since the early twentieth century, the Claretian Order was a Spanish order with primarily Spanish-speaking priests.[55] Expelled from Mexico during the revolution, the Claretians were intimately aware of the revolutionary and post-revolutionary anti-Catholicism in Mexico. The links between the order, the pre-revolutionary Mexican government, and Spain were not lost on Mexican immigrants who supported the role of the Church before the revolution and those who did not. Anti–Catholic Church Mexicans, those who supported the revolution, linked these priests to Spanish colonialism and Mexican dictator Porfirio Diaz. Those who supported the Church considered the Claretians symbols of perseverance and hope.

In 1925, Claretian Father James Tort, a Spaniard, took over responsibility of Our Lady of Guadalupe. He had served in the Canary Islands and then Mexico City, from where he fled during the Catholic purges of 1914. He then served in Arizona and Texas before coming to Chicago and quickly became popular with Mexican and Irish Catholics in the neighborhood. People around him described him as tireless and having a "magnetic personality"—important traits for someone responsible for religious instruction not only in South Chicago, but also in the boxcar camps and several nearby cities including Milwaukee and Waukegan. Tort was responsible for raising enough money from non-Mexican Catholics (primarily Irish) to build the large Our Lady of Guadalupe on South Brandon Avenue and 91st Street. This new building quickly became a proud symbol for Mexican Catholics in South Chicago.[56] The Cordi-Marian Sisters, a Mexican order that had also been expelled from Mexico, joined the Claretians in 1927.

Despite Tort's success, Catholic resources were not anywhere close to that of the Protestant missions in the area. This trend was not unique to Mexican South Chicago. According to McCarthy, Catholic leaders were "hesitant to propose programs to maintain Mexican Catholic identity in Protestant America." It was not until 1930 that Catholic social services noticeably

Fig. 5.2. First Our Lady of Guadalupe Catholic Church in South Chicago. Photograph courtesy of the Southeast Chicago Historical Society (neg. no. 81-102-2).

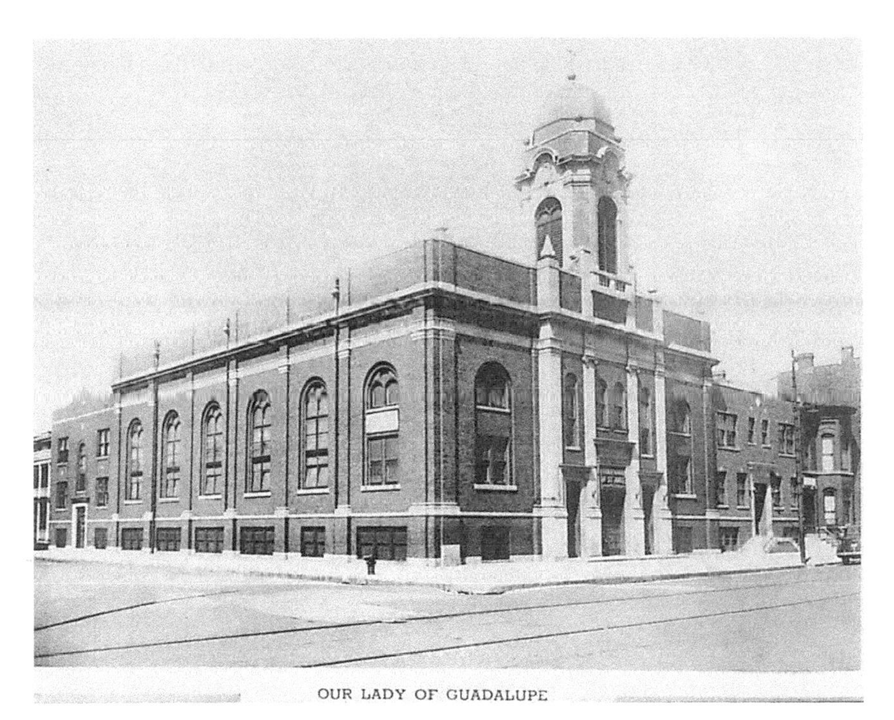

OUR LADY OF GUADALUPE

Fig. 5.3. New and current Our Lady of Guadalupe Catholic Church in South Chicago. Photograph courtesy of the Southeast Chicago Historical Society (neg. no. 81-28-3).

expanded into Mexican immigrant neighborhoods as Catholic bishops "adopted Protestant social welfare strategies to reclaim their lost sheep."[57]

Regardless of religious inclinations, Mexican husbands and parents of young women were anxious about the entrance of wives and daughters into the commercial or industrial workplace because of their increased independence that, in the eyes of husbands and parents, would lead to a greater opportunity for social and sexual autonomy. Entry into the workplace and participation in unsupervised social activities undermined patriarchal and familial control by creating a youth culture wherein daughters explored romantic relationships away from the "watchful eye" of parents and relatives.[58] Mexican parents' fear of negative influences on their daughters went beyond the workplace.

Parents attempted to control interaction between girls and boys in schools and during leisure activities. Socorro Zaragoza, an elementary school student in nearby Gary, Indiana, complained that her father would not let her do anything because "he is afraid I will be like Americans." She protested that although—or perhaps because—her father had been in the Chicago area for sixteen years, he did not want her hanging around American girls because they "go out alone, talk back to their parents and don't help their mothers." She then reasoned that her father was right because he had seen how "American girls treat their mothers."[59] Furthermore, Zaragoza's father excused her from gymnasium class because he believed it was immoral and only let her attend movies or social gatherings two or three times a year when he accompanied her.[60]

As with other ethnic groups in Chicago, *Mexicanos* in South Chicago discouraged ethnic and racial intermarriage. Ignacio Vallarta, interviewed while a grade-school student in Northwest Indiana in 1928, emphasized that "The Mexicans hate to see other Mexicans marry colored people," despite the fact he considered African Americans "human beings, the same as everybody." Vallarta argued that it looked "bad among his people" because there were "plenty of Mexicans to marry." He went on to mention that "some of the Mexicans who marry colored say it's because there aren't many Mexican girls and they don't like the Mexican girls who are here. The Mexican men here say that the Mexican girls of the United States have too much liberty. They want to marry Mexicans newly from Mexico. Mexican girls here want Mexican men who have been in the United States."[61] According to his description, men as well as women had expectations of the types of Mexican immigrants they would marry.

Lacy Simms, a South Chicago Protestant pastor, considered the increasing levels of intermarriage a positive development for the Mexican community in general. He argued that intermarriage accelerated the Americanization

Fig. 5.4. Wedding photograph of Justino Cordero and Caroline Kon. Photograph courtesy of the Southeast Chicago Historical Society (neg. no 81-120-38).

process. "I have seen happy marriages of negroes and Mexicans," he said. "The Mexicans are marrying Poles of the second generation and some Jewish, Italians, Germans, etc. Those who have inter-married often become citizens. Some Mexicans do become citizens now."[62] Assimilationists viewed the low naturalization rates among Mexican immigrants a social and economic problem for the community, and Simms believed that the benefits of intermarriage outweighed any community stigma.

Francisco Huerta, the newspaperman, remarked that "many Polish girls marry Mexicans at 45th and Ashland. The Polish men wanted to kill the Mexicans for going with Polish girls." Tired of being considered less than white, "the Poles," according to Huerta, "say they don't want to mix white with yellow blood; but they don't know that our skin is dark but our blood is white."[63] Notwithstanding the Mexican community's general disapproval of European-Mexican intermarriage, Huerta objected to ethnic Europeans' resistance to intermarriage. Huerta's desire to be classified as white underscores the prominent hierarchical structure in Chicago that classified migrants by color and provided more opportunity to those considered white, regardless of ethnic heritage.

In discussing interracial marriage and miscegenation laws, historian Peggy Pascoe argues that "the claim that interracial marriage was unnatural drew much of its power from cultural assumptions about the nature of race."[64] It was this racial hierarchy and the cultural expectations of many *Mexicanos* in the South Chicago area that together pressured members of the community to conform. Although anti-miscegenation laws did not play a role for Mexicans in South Chicago, this disapproval of miscegenation was about advancing up the color line. Community leaders of ethnic groups who considered themselves above Mexicans on the color line had the same objections to a member of their group marrying a Mexican. In other words, this cultural racial norm meant that Polish parents did not want their kids marrying Mexicans, and Mexican parents did not want their kids marrying African Americans. In explaining why these ideas were—and in many places still are—so popular, Pascoe argues that "American racial systems have always worked most effectively when they are taken so completely for granted that their structures are more or less invisible to Whites."[65] Furthermore, as Pascoe observes, "Every successive American racial regime, beginning with slavery, but continuing with the taking of Indian lands, the establishment of segregation, and the development of American immigration restrictions, expended a great deal of energy making its racial notions appear so natural that they could not be comprehended as contradictions to a society ostensibly based on equality."[66] These racial notions, which were not new to many immigrants, motivated the desire to separate and segregate those darker than themselves. Intermarriage, it was feared, would cost the "whiter" group status and respect in hegemonic white America.

Mexicanos in the community wanted to be considered white but did not agree on what it took to be considered white by others. They disagreed with the dominant Anglo-American community that whiteness required naturalization. The use of English, though, remained more complicated. Although Mexican men believed that allowing women to learn English was a stepping stone to Americanization, the idea of their learning English themselves was less contentious but still controversial. Some South Chicago Mexicans frowned on the use of English within the community because to them it signified a rejection of cultural pride and a willful personal distancing from Mexico. On the other hand, many in the community recognized that speaking English outside of the community could improve individual and collective Mexican status in Chicago. One complexity of this language debate was that the divide between Spanish and English sometimes weakened in practical usage. It was not a clear-cut matter of acceptance or rejection of English. Since all languages evolve, Anglicisms crept into the spoken language of Mexican Chicago. In 1928, Mexican anthropologist Manuel Gamio listed twenty-six words used and accepted in everyday "street" Spanish in Chicago that adapted English words into a Spanish structure. Thus, the Spanish "patios" became "yardas," from "yard"; a boarding house was a "casa de borde"; "casita" became "chante" for "shanty." Instead of or in addition to asking "què pasa?" Mexicans in Chicago might jokingly inquire of a companion, "Guasumara?" from "What's the matter?" A favorite treat was "aiscrin" rather than "nieve." Other Anglicisms came from men's work activity: "dipo" was a depot, "estoque yardas" were "stockyards," and "bos" was "boss." It is difficult to know how much of the word choice was based on frequent activity and how much was the result of Gamio's own interests and assumptions about his informants. His list also included "punchar," which might indicate the rough, physical nature of men's encounters in their daily lives, but could also be more a representation of Gamio's view of the men as prone to fighting.[67] These language adaptations documented by Gamio are examples of Mexican Chicago's "third space," the unique space between being a Mexican in Mexico and being a white "American."

Catholic and Protestant churches differed on Americanization in both principle and practice. Protestant churches and most settlement houses advocated for full assimilation and a cultural immersion that included exclusive loyalty to the United States, conversion to a Protestant faith, a full embrace of U.S. democratic values and "well-developed social habits." The

Catholic Church, on the other hand, preferred a more gradual Americanization that encouraged immigrants to embrace U.S. democratic values while keeping parts of their ethnic culture.[68]

As highlighted in the *Daily News* article and as part of a process of negotiation, some members of the community took advantage of English classes. They did this, however, aware of the cultural and social risks posed by speaking English within the community. Within the Mexican community of South Chicago, language assimilation was a problematic issue linked politically to the loss of ethnic culture and therefore to the betrayal of one's Mexican cultural identity. Some saw learning English as the first step down a slippery slope that would encourage individuals to abandon the Spanish language of their community. An editorial in the Chicago Spanish-language newspaper *Mexico*, probably written by the newspaper's owner Francisco Huerta, approved of the ease with which Mexican children explained themselves in English, singing and talking to their friends. But he was bothered that most of them "are completely Americanized. It is easier for them to say daddy than *papa*, block instead of *manzana*; I don't care for *no me importa*, they sing 'Follow the Swallow' and they forget the words to 'Sobre las Olas.'"[69] Lack of Spanish, Huerta contended, created a youth ignorant about Mexico's culture, geography, and history. Not only did these children not honor the Mexican flag, they barely knew they were Mexican, and had only heard so from people besides their parents. According to Huerta, parents had to make sure their children knew and took pride in the language and history of Mexico. If not, Mexico would lose their children entirely to English and Americanization.[70]

While assimilationists urged Mexicans to learn English as they had urged every other immigrant group, Mexican community leaders acknowledged that the acquisition and use of English could also be employed to combat discrimination against individuals and the community. Other Chicago Spanish-language newspaper editors (and by proxy much of the Mexican middle class in Chicago) took a different tack in explaining the relationship between language and Mexican identity. They tried to dispel criticism that the study of English led to the loss of one's Mexican identity or somehow betrayed the Mexican homeland by linking the need to learn English to Mexican patriotic holidays, Mexican patriotism, and the responsibility to return to Mexico. In an editorial published in the *Correo Mexicano* on September 16, 1926—considered by many Mexicans their most important patriotic holiday—the writer highlights the significance and symbolism of the holiday in order to reach Chicago-area Mexican readers, including those in South Chicago. The author of the editorial argued that Mexicans throughout the area needed to learn English (and honor those who taught the community English) in order to better themselves as laborers, skilled workers, or students. In going beyond the argument that learning English would benefit members of the community and their families, the author drew on individual patriotism and the sojourner attitude within the community to advocate the learning of English in order to take new skills and knowledge back to Mexico.[71] By linking his pro-English comments to Mexican patriotism and the importance of the eventual return of Chicago-area Mexican immigrants to Mexico, the editorial's author promoted the learning of English as pro-Mexico and beneficial to Mexican culture both in Chicago and in Mexico. Beyond that, the learning of English and the vocational and intellectual skills workers could acquire after learning English would benefit Mexico on the immigrants' return.

Assimilation started with education. The board of education, settlement houses, churches, and immigrant advocacy organizations initiated the assimilation process through the formal education of immigrants, including teaching English to adults. In Chicago, this English-language education of Mexican immigrants was conducted and filtered by assimilationists who were, for the most part, well intentioned. Although there were various—and at times competing—reasons for assimilation, many Mexicans agreed that learning English would improve their social contacts and economic conditions. Paul S. Taylor argued that "a lack of knowledge of English was a material loss recognized and experienced by many, which handicapped them in securing employment and promotion." At the same time, *Mexicanos* realized that learning English was not a golden key to success or acceptance. One of Taylor's informants argued that being able to speak English was no guarantee of avoiding discrimination. He complained that "There are many Mexicans who speak English, but even they do not get ahead."[72] According to this community member, many Mexicans did choose to learn English and pursue assimilation as a way to improve their ability to succeed economically. If they did not so choose, they risked being coerced into Americanization by external assimilationist pressures that included further economic and social exclusion at the workplace and in the local non-Mexican community. Community members' ability to negotiate the use of Spanish or English within and outside of the community remained a critical component in the creation of a supportive environment.[73]

Debates within the community of this key paradox directly influenced the community's cohesion and development. Internal divisions affected attendance of English classes. Many husbands did not want their wives to learn English because they believed it would make women too independent.[74] For men, refusing to let the women in their families attend English classes was a way to retain Mexican cultural mores and maintain closer control of women in the face of rising pressure from social workers and others to adopt "American" gender norms.[75] Community organizers countered the perception that learning English was a threat by sponsoring English classes through Mexican mutual aid societies. By having internal groups conduct English classes, organizers hoped to alleviate fears that linked the learning of English to frequently condescending and sometimes discriminatory outside agencies. In 1925, the *Sociedad Mutuo-Recreativo Benito Juarez* advertised their free English classes that were taught by professors from the University of Chicago. By offering these classes, mutual aid societies expanded their services and value to the community and demonstrated internal resolve to provide an opportunity to learn English without the accompanying push to abandon parts of Mexican culture that was a central part of Board of Education and Settlement House curriculum.[76]

Reasons for learning or not learning English were more complicated than just being about culture and economics. Not knowing English had more immediate consequences when dealing with police and other officials. Non-English speakers who interacted with the police or court system were more likely to be wrongly convicted and imprisoned than English speaking *Mexicanos*. Agustin Fink expanded on this during a 1926 interview with Manuel Gamio. Fink, born to a German father and Mexican mother, was raised and educated in Mexico. He was trained as a civil engineer before coming to Chicago in 1922. By the time of his 1926 interview with Gamio, he was selling insurance in Chicago for Aetna. He reported that many Mexicans "got into lots of difficulty with the law and the police." The Chicago court system did not have Spanish-speaking lawyers, so according to Fink, judges convicted and sentenced Mexican defendants because they could not understand the Spanish speakers. Fink believed that the police had other motives for targeting Mexicans. The police, Fink argued, "accused innocent people to increase their arrest rate or because of baseless suspicions." Despite his middle-class status and life outside of the working class neighborhoods, Fink's perception of the discriminatory treatment of Mexicans who did not speak English was accurate.[77]

Dr. Juan Medina, a Jalisco-born dentist who first migrated to Chicago in 1914 and occasionally worked for the Mexican consulate, described police discrimination in similar terms. The consulate had to work to protect Mexicans from "the police who often accuse them because of a lack of interpreters and because they assign guilt without sufficient proof." Asserting that "even a negro has more protections because they [*sic*] speak the language," he made the case that Mexicans who did not speak English were even worse off in the hands of the police than African Americans who generally occupied a lower position in the racial hierarchy of the dominant society. However, the consulate could protect only those individuals who had entered legally and registered with them, thereby leaving many immigrants to turn to each other for support.[78]

In addition to the legal restrictions placed on the Mexican consulate's ability to protect immigrants, the fact that the consul's office in Chicago was responsible for a large swath of the Midwest, including parts or all of Illinois, Wisconsin, Indiana, Minnesota, and Iowa, meant that Mexicans had little recourse when they ran into legal problems—very little, in fact, when considering that in 1940, the Chicago consulate operated with a total paid staff of seven workers.[79]

Even if the police did not, as a matter of course, consciously accuse Mexicans whom they knew to be innocent, the perception that they did by Mexican elites like Medina and Fink is significant in itself. The belief that governmental institutions like the court system and the police did discriminate reinforced Mexicans' sense of being marginalized and made them less likely to go outside their ethnic community to look for help. Many Chicago police officers were Polish, and a comparatively uneven arrest rate of Mexicans and Poles involved in violent disputes across Chicago in 1926 and 1927 only furthered Mexicans' perception of corruption and discrimination in the police force.[80]

Since many Mexican immigrants to South Chicago planned to return to Mexico, they did not actively seek to Americanize or naturalize. Those Mexicans who took advantage of English classes in order to improve employment prospects nonetheless understood Spanish was a critical component of their cultural and political identity. They believed themselves to be sojourners who were escaping political turmoil and would eventually return to Mexico in a better economic position than when they had left. The physical and geographical circumstances of South Chicago provided its Mexican community with prospects that differed from those in rural communities in Mexico and the other Mexican enclaves of the Chicago area. Moving into a pre-existing ethnic working-class neighborhood in a crowded industrial environment had drawbacks when compared to rural life or life in the urban American Southwest, yet living in a densely populated neighborhood with a large Mexican population allowed for community-building opportunities. These opportunities, however, were gendered. Cultural limitations imposed on women and girls made it difficult for them to take advantage of many of the opportunities available in a large urban area.

6

Resistance

On June 2, 1928, José Vasconcelos, the prominent Mexican scholar and political activist, spoke passionately to a crowded Bowen Hall in Jane Addams' Hull House on Chicago's Near West Side. Gathered under the auspices of the mutual aid society Ignacio Zaragoza, Mexicans from all over the Chicago area heard Vasconcelos speak of corruption in the Mexican government and of rampant nepotism at the highest levels. Vasconcelos pleaded to those assembled in the hall to "never forget or cease to show interest in our country and in the land in which we first saw the light of day." Addressing the reasons many of those in the hall came to the United States and appealing to their strong Christian cultural identification, Vasconcelos implored those present to keep their interest in Mexican affairs:

> For, if we are working hard and suffering, it will not always be so. We are but the children of Israel who are passing through our Egypt here in the United States doing the onerous labors, swallowing our pride, bracing up under the indignities heaped upon us here. If we expect to return and to escape all this, as all good Mexicans ought to, then we should show our interests in the affairs of our country, from this Egypt of ours.[1]

Vasconcelos used these biblical references comparing the experiences of Mexicans in Chicago to those of the Israelites as slaves in Egypt to explain the suffering, harassment and discrimination endured by Mexicans at the hands of white and ethnic European Chicagoans. He appealed directly to a Mexican sense of cultural responsibility and duty to persevere for themselves, for their family, and for their *patria,* or homeland. For them, he was saying, the United States was not a promised land but its prelude: "good Mexicans" ought to "expect to return" to Mexico "and to escape" the conditions they faced in Chicago. In his carefully crafted speech, Vasconcelos implored Mexicans not only to work hard while keeping their Mexican culture and celebrations, but also to keep an acute interest in the post-revolutionary political and economic well-being of Mexico.[2]

José Vasconcelos' presentation is significant in several respects. Vasconcelos, a writer, philosopher, and politician, was an influential yet controversial Mexican national figure who created the modern Mexican educational system. He is best known today for *La Raza Cósmica*, in which he argued that the mixing of races was natural and desirable for a prosperous society and that ethnic variety was valuable and beneficial for humankind. He had served as head of the National Autonomous University of Mexico (UNAM), created the Secretariat of Public Education (SEP), and served as the secretary of public education. He came to Chicago in 1928 as a candidate for president of Mexico. Although he lost the 1929 election to Pascual Ortiz, his visit to Chicago signaled the importance and influence of the local Mexican immigrant community barely a decade into its existence. Vasconcelos, and other politicians who came to Chicago, depended on the residents who sent money home to also send word about which candidates to support and vote for. Part of the widespread sojourner attitude was a continued interest in Mexican affairs and in improving conditions in Mexico for their eventual return. Pascual Ortiz's visit to Chicago came only two months after his triumph over Vasconcelos.

Although economic and political variables, as well as high-profile Mexican visitors to Chicago, played roles in how Mexicans in South Chicago shaped their cultural perception of their homeland, they nonetheless used the social and political legacies of what many of them considered to be an ongoing revolution in Mexico to shape their self-identity and their ideas of Mexican culture. The community's resistance through acting, reacting, and organizing against the ever-present harassment, discrimination, and hardship in Chicago, their sense of cultural pride and obligation to Mexico, and the political and economic conditions in their *patria* were all factors in creating a physically and culturally strong Mexican community in South Chicago—a community able to resist assimilationist pressures and negotiate on its own behalf.

In this chapter, I focus on various acts of negotiation and resistance in the everyday lives of Mexicans in South Chicago and the surrounding Mexican communities. I argue that the members of the Mexican community in the Chicago area developed and used forms of resistance in hopes of finding ways to improve their everyday life without compromising their sense of pride for Mexico or the facets of their Mexican cultural heritage they held close to their hearts. Along with individual acts of resistance—such as the language debate highlighted in the previous chapter that emphasized the learning of English only to the extent necessary to advance economically while limiting its use to communication with non-Mexicans in the workplace—members of the Mexican community formed mutual aid societies, pro-*patria* clubs, social clubs, and athletic teams. In forming these organizations, *Mexicanos* in and around South Chicago sought to reinforce a sense of Mexican cultural solidarity, while simultaneously providing social and economic support for members of their community.

While complicating our understanding of the Mexican community of South Chicago, as well as of individual Mexicans who were active members of the organizations within the community, we must acknowledge that Mexicans were neither entirely victims of various kinds of oppression in society, nor were they solely focused on monetary gain. It is important to recognize the role traditional labor and working-class organizations played in everyday life and then expand beyond those institutions to examine the people themselves. This chapter analyzes the ways in which Mexicans in Chicago resisted discrimination from those around them to create a community that, despite its divisions, provided support and reinforced pride in their *patria*.

A significant cultural difference between Mexican immigrants and most other South Chicago immigrant groups was that most Mexicans planned to return to their *patria*. As Jesse Parez, a Mexican living in Chicago, put it, "Returning to Mexico is much simpler than returning to Europe, so our group quite naturally are [*sic*] reluctant about becoming citizens when we face so much discrimination here, and know that Mexico is not too far to return to."[3] Many, if not most, of the Mexican immigrants in South Chicago saw themselves as sojourners who were escaping political turmoil and planned to make enough money to return to Mexico in a better economic position than when they had left. Most immigrants did not set a firm date to return to Mexico. Instead, they waited for a "right time" that in some cases never came. Although not living in Chicago, Poet Miguel Arce voiced the attitude of most Mexican immigrants in the United States. He hoped "to go back to Mexico when the situation is settled there, perhaps I can do more in my country and however that may be I would be better there."[4]

Many did return to Mexico. Others who never returned to Mexico nevertheless kept alive the dream of returning someday. In 1926, three young men in South Chicago, workers at a steel mill, expressed the desire return to Mexico "when it gets quieter there" even if that meant making less money. Max Guzman agreed. The "average mentality of the Mexican steelworker" was to work and prepare to go back once things settled in Mexico. He declared, "The old saying that everybody was going to go back to Mexico, that is the main thing you see," even when "people would live" in Chicago "for years and years."[5]

The concept of Mexicans as sojourners, of eventually returning to Mexico in a better economic condition and as an asset to Mexico, operated on two levels. First was the effect on Mexicans themselves, the choices they made, and the rhetoric they used to describe their experiences. Second was how the sojourner attitude affected the way non-Mexicans—city officials, social workers and reformers, European immigrants, and the federal government—perceived and treated Mexicans. A shared expectation of returning to Mexico created a common bond that helped solidify a sense of community despite the fact that most did not plan to be in South Chicago for long. Mexicans changed their environment to improve their living conditions and sense of belonging despite this sojourner attitude. However, Mexicans' expectation of eventual mobility did intensify anti-Mexican sentiment locally and nationwide, as outside groups equated low levels of naturalization resulting, in part, from the sojourner attitude with hostility towards the United States. When Esperanza González, a girl acting as an interpreter for her parents in a meeting with a Chicago-area United Charities social worker, was informed by the social worker that a neighbor had taken out naturalization papers in order to receive a mother's pension, she was shocked and replied, "What! Did Mary Díaz change flags?"[6] Arce's and González's comments reveal the external pressures on Mexicans to naturalize, expectations from within the community that one would not seek naturalization, and an internal sense of pride in their Mexican citizenship and identity.

The idea of a "continental citizenship"[7] that included pride in Mexico while concurrently acting in the best interest of the United States made sense to many in the Mexican community as they saw few positive reasons to "change flags." Having been born on the American continent, Mexicans objected to United States monopolization of the term "American." Arce's brother, Miguel, declared, "I believe in and love Mexico and the rest of Latin America although I don't believe that my father-land is limited to Mexico."[8] Mexicans were already Americans and United States citizenship was of little day-to-day economic or cultural benefit. Many Mexicans in South Chicago not only found naturalization unnecessary and of no practical value in everyday life, but also found the idea of renouncing Mexican citizenship offensive and contradictory to their cultural mores.

Historian Gabriela Arredondo points out that "Mexicans in Chicago were not simply another ethnic group working to be assimilated into a city with a long history of incorporating newcomers." Because of the discrimination all around them, "Mexicans fought the current of pejorative qualities ascribed to 'being Mexican' by constructing their own Mexicanness to battle anti-Mexican prejudices."[9] Arredondo's argument holds true for South Chicago, where Mexicans looked to maintain their nuanced Mexican South Chicago identity instead of assimilating. Although this identity created resistance to full assimilation to "American" culture, Mexicans still wanted to be racialized as white, rather than being relegated to the less-than-white status occupied by most working-class *Mexicanos.*

Lacy Simms, the South Chicago pastor introduced in the last chapter, was a Texas-born Congregational minister who came to Chicago after missionary work in northern New Mexico. He believed there were five primary reasons why South Chicago Mexicans did not naturalize in any significant numbers. First, Simms argued, Mexicans did not seek citizenship because of their migrant lifestyle. He believed that even if Mexicans wanted to become citizens, "their moving about the country is so great that no two witnesses in any one place can swear that they have known them for five years," as required by law. Second, Simms believed that Mexicans had never "been citizens of any country in any real sense of participating in their government" because of the constant revolution and lack of democratic participation during the Porfirio Díaz regime. Third, he blamed the lack of naturalization on Mexican resentment of the United States because of "past military conflicts" between Mexico and the United States. Fourth, Simms reasoned that Mexicans avoided naturalization because they did not feel welcome by Americans. His fifth reason was that they "carry the thought of returning to their native country." Although many Mexicans held feelings of resentment for the United States government because of U.S. invasions of Mexico, the last two points resonated the most strongly within the Mexican community of South Chicago.[10]

The widespread idea of being a sojourner helps to explain the extremely low naturalization rates of Mexican immigrants in South Chicago and the rest of the United States. Indeed, Mexican government officials discouraged naturalization and encouraged parents to retain Mexican citizenship for their U.S.-born children.[11] Precise percentages are difficult to come by; scholars cite unquantified "low naturalization rates" for the interwar period. Using nationwide data, UCLA economist Leo Grebler gives the number of Mexicans naturalized in four-year blocks and the percentage of Mexicans relative to all aliens naturalized during that period. While the total Mexican-born population in the United States was 4.5% of the entire foreign-born population in the United States in 1930, Mexicans comprised only 0.1% (529 individuals) of all those naturalized from 1925 to 1929. During the 1930–1934 period, 1,226 Mexicans naturalized nationwide, making up 0.2% of the total number of those naturalized. Because these are national numbers, they are not very useful for determining regional variations.[12] The closest we come to naturalization numbers in the Chicago area during the 1920s and early 1930s are those provided by Paul S. Taylor. Using his personal observation around 1930 and employing "not entirely comparable statistics" of 1,800 Mexican workers in three industrial plants in the Chicago area, Taylor posits that no more than 2% of Mexicans in the Chicago area were naturalized.[13] An additional factor to consider when examining all of this quantitative data is the bureaucratic lag time. Because of the variable lag time in the multiple stages of the application process for naturalization, there is no way to know the number of Mexicans in the United States meeting the minimum time requirement for naturalization eligibility in any given year compared to the total number of Mexicans (documented or undocumented) in the country.

The low naturalization rates fueled anti-Mexican feelings in the larger South Chicago community. The late 1920s and early 1930s was a time of increasing economic and social pressure on Mexicans to prove their loyalty to the United States by assimilating or naturalizing. Although such pressure was not isolated to South Chicago, it was present in the neighborhood. Not all those who had studied the "Mexican problem" agreed that Mexicans were a problem. In a 1931 analysis of Mexicans in Chicago written for a commission of the Chicago Church Federation, Robert C. Jones and Louis R. Wilson asked a seemingly simple question: "What place will we make for the Mexican in our American life?"[14] They followed the question by adding that Mexicans "undoubtedly" had a contribution to make in Chicago. Their report reacted to the general assumption that naturalization was necessary to true citizenship. It was an attempt to convince the Comity Committee of the Chicago Church Federation to invest time, energy, and money in improving current and/or creating new Protestant Churches to serve the needs of Chicago Mexicans because of their importance and loyalty to the city. They did this at a time when the depression-era nativist backlash and subsequent repatriation campaigns served to solidify Mexican resistance to assimilation and naturalization in Chicago. While Jones and Wilson articulated a concept of citizenship that was atypical for Chicago at the time and so was not indicative of prevailing attitudes, their assertion that naturalization was not a prerequisite for acts of citizenship highlights the core question of this citizenship and whether naturalization was a prerequisite for being a contributing member of the larger society.

In the observations that followed, Jones and Wilson argued adamantly for inclusion and acceptance of Mexicans within Chicago, asserting that Mexican immigrants brought with them "a large capacity for useful citizenship."[15]

Separating the concept of citizenship from any necessity to naturalize, the authors broadened the concept of citizenship to include those who gave their labor and skills to the economic and cultural betterment of the country without regard to their immigration status, The contents of the Church Federation report echoed several of the points made by José Vasconcelos during his speech at Hull House. Both addressed reasons for low naturalization and highlighted Mexicans' hard work. Jones and Wilson emphasized the hard-working nature of Mexican immigrants while explaining their low naturalization rates as being a result of Mexicans' feeling that they had little to gain through a change in citizenship. Vasconcelos had urged Mexicans in Chicago to work hard despite their suffering while "passing through [their] Egypt" so they could persevere and do well for themselves and the good of their native Mexico.[16] Vasconcelos placed more importance on working hard for the long term benefit of Mexico, but the overlap demonstrates shared insider and outsider perceptions of issues confronting Mexicans. As one South Chicago Mexican observed, "Until the Latin American's idea of the United States and the American's idea of Mexicans changes I don't believe that there will be many Mexicans seeking to become American citizens." The immigrant argued that these ideas transcended generations and were about respect: "Even those young Mexicans who think that they are superior to others because they have adopted American ways are still Mexicans and if any American makes a slighting remark about Mexico he still feels himself insulted." It was, in the end, about pride in one's culture and the resentment against U.S. intervention in Latin America. The image of the United States as the "octopus of the north" needed to go in order for hearts to start changing.

In attempting to lay a foundation for the popular acceptance of Mexican immigrants into Chicago society by defining naturalization as an unnecessary prerequisite for a person to perform an act of citizenship, Jones and Wilson offered three reasons why Mexican immigrants seldom renounced their Mexican citizenship for that of the United States. First, the sociologists pointed to a "strong feeling for his native land and a sense of racial solidarity." Second, they invoked a continental concept of the Americas on behalf of the universal male Mexican, stating that the immigrant knew that "he too, is an American and proud of that fact" and remained aware of his proximity to "the border of his own land." Third, they pointed out the discrepancy between Chicago's acceptance of Mexicans' "labor at unpleasant tasks" but rejection of their culture and society.[17]

Even while Jones and Wilson failed to discuss the important roles played by women in agricultural and industrial environments and use "the Mexican" as representative of the entire working-class Mexican population, they made an argument for the acceptance of all Mexicans based on their agricultural and industrial work. Moreover, they argued that Mexicans had a "mechanical aptitude" which the dominant society should employ, rather than offering "ill treatment and social disapproval." In further defense of the Mexicans in Chicago—including South Chicago—and to the value of Mexican immigrants to the greater area, Jones and Wilson went on to describe "the Mexican" as one who had inclinations towards the "mystical" and was open to "religious expression" with a strong sense of "artistic beauty."[18] Although many of their characterizations were stereotypical, Jones and Wilson's sympathetic portrayal was aimed at convincing their readers in the Chicago Church Federation that Mexican immigrants could and did play vital roles in the economic and cultural prosperity of South Chicago and the other Mexican communities of Chicago. This report was significant because the Chicago Church Federation consisted of all major Protestant denominations in the area, including those operating in South Chicago. One function of the federation's Comity Committee was to determine the need for new churches and church missions within the city—including the Mexican community of South Chicago—and to regulate the number of new Protestant churches and missions that attempted to operate in the area. In attempting to counter the anti-Mexican, nativist rhetoric common within the larger society, Jones and Wilson hoped to promote the importance and loyalty of Mexicans in Chicago and encourage more Protestant church activity and support within the Mexican community of South Chicago.

Attempts to regulate Protestant churches within a community were not always successful. A Mexican immigrant in South Chicago identified only as J.O.V. complained about the Mexican churches in the neighborhood: "The trouble in South Chicago is not between the members but between the pastors." Pastors at two of the Spanish-speaking "Mexican" Protestant churches in the neighborhood spent too much time poaching members from each other's congregation, he said. "Both of the pastors were Latin-Americans and each wanted to have the biggest church and to that end tried to take the members away from each other." J.O.V acknowledged that it was not entirely the fault of the pastors. "The fault partly lay with the mission boards which judge the value of the work according to the number of people who attend. The pastors have to send in their reports and the superintendents make their visits so that rather than trying to develop the religious life of their people they spend time filling up the church." In a system where more members meant more money from mission boards, churches focused on enrolling members in order to continue to receive operating money. J.O.V, who is identified as a Protestant who had converted in Mexico, makes several valid criticisms that need to be weighed against the denominations' priorities to focus on Americanizing and converting working-class Mexican Catholics. As with the construction of the new church of Our Lady of Guadalupe, outside money was necessary to supporting local religious infrastructure.[19]

Research notes by Robert Jones produced circa 1930 list six churches in South Chicago. Bird Memorial, referred to by Jones as United Evangelical Mexican Church, was located at 9135 Brandon Avenue. First Mexican Baptist Church of South Chicago was at 90th Street and Houston Avenue. The Baptist South Chicago Neighborhood House was in the Bush at 8514 South Buffalo Avenue, and a "Spiritualist Center" was near the corner of South Burley Avenue and 90th Street. Also listed is Our Lady of Guadalupe Catholic Church at 91st Street and South Brandon Avenue.[20]

Maintaining Mexican citizenship was a seemingly contradictory way for immigrants to strengthen community in South Chicago and to create an environmentally distinct Mexican South Chicago. National ties to Mexico reinforced a unifying cultural concept. Tensions over Mexicans' low naturalization rates were arguments over whose citizens they were. As the number of Mexicans in South Chicago continued to grow before the Great Depression and Mexicans physically and culturally established themselves in the neighborhood, members of the more dominant culture increased the pressure and intensity of their rhetoric around Mexican immigrants Americanizing. The report by Jones and Wilson for the Chicago Church Federation did not represent the majority view. Its publication was limited and aimed primarily at the Church Federation, other religious leaders, social workers, and immigrant advocates.

Anti-Mexican sentiment was high in South Chicago among other ethnic immigrants especially the Polish and Irish. In 1928, a group of Chicago-born "Polish Boys" described Mexicans as "very dirty" and said that they "spoil the new houses they occupy" because they crammed too many people into one house and lived "like rats all cooped up." Pointing out one particular house on 83rd Street and Burley Avenue in the Bush section of South Chicago, one of the boys declared, "You could smell the place when you passed in front of it on a dark night."[21] According to the 1930 U.S. census, the only building near that intersection with Mexican residents was 8260 South Burley Avenue. This building, on the northwest corner of the intersection, was home to Jesus Cerda, his wife Margarita, his three children, and his brother Francisco. Lost on the Polish boys was the fact that the two other apartments in the building were home to Polish immigrants: one was occupied by a young Polish couple, the other by a large Polish family of seven. In 1928, Mexicans still stood out in the Bush even as they started to take over several blocks in the Millgate section. The few Mexican families who lived in the Bush were scattered throughout the section, with most blocks having no more than two Mexican-occupied apartments. The kids' vivid description of Mexicans spoiling their local environment and adding to the pollution of an already environmentally compromised area was one repeated by ethnic European adults and officials alike. Describing newer immigrants groups as dirty was a common disparagement—and one that bypassed the fact that these immigrants were confined to the worst, most decrepit housing in the worst parts of the neighborhood.[22]

Mexican community leaders were acutely aware that this was not an insult reserved for use by kids or neighborhood ethnic rivals. The author of an October 1928 *Mexico* newspaper editorial wrote passionately against public comments by Dr. Benjamin Goldberg, head of the Municipal Tuberculosis Sanitarium of Chicago. Although the comments were part of a paper read at a professional conference being held in Chicago, they were picked up and printed by the *Chicago Tribune*. The *Tribune* reporting gave readers the impression that this was an official position held by city health officials. The *Mexico* editorial writer called the article "a summary of insults by Dr. Goldberg" that were "inspired by hate and malice and so contrary to truth" that they endangered "good relations between the Mexican and the American People."[23] The author goes on to call Dr. Goldberg a "quack" who selected the "already ill-treated Mexican" as his target while saying "nothing of other nationalities."[24] Being even more specific, the author blasted Goldberg because he "does not speak of Gypsies, Italians, or Polish. It is the Mexican who comes here to infect them and to steal their money."[25] Goldberg's comments, likely influenced by the popular eugenics movement of the period, angered Mexicans not only because he put Mexicans in a negative light, but because he implied that Mexicans were predisposed to be sick and, according to Goldberg, "a menace" that needed to be controlled and contained—much like tuberculosis itself—for the sake of the general public.[26]

The *Mexico* editorial quotes arguably the most inflammatory statement credited to Goldberg: "The Mexican emigrants who cross the Rio Grande in numbers and who multiply very rapidly are not only undesirable but they also constitute a menace to the health of the American people."[27] Goldberg believed that they constituted a menace to Americans—and here he meant "white" Americans—because of the higher-than-average infection rate in Chicago's Mexican community and the stereotype that they had more children than other immigrant groups. The study, which does not examine closely other ethnic groups, also fails to take into account the environmental racism, poor housing, and poor sanitation experienced at a much higher degree by the Mexican community than by other ethnic groups.[28]

Goldberg's paper, which he read before the Vital Statistics Section of the American Public Health Association's Annual Meeting in Chicago on October 18, 1928, was published four months later in *The American Journal of Public Health and the Nation's Health*. In a response to the paper, Dr. Godias Drolet, a statistician for the research department (and later associate director) of the New York Tuberculosis and Research Service, criticized Goldberg for his poor methodology and personal biases. Drolet introduced the idea that environment is an important factor, arguing that "tuberculosis susceptibility or resistance can, under similar circumstances, be developed equally." Drolet also went on to call out Goldberg on his eugenicist beliefs: "Personally, I do not like the phrase about the Mexican that he is of another race," he protested, and then gave a quick history lesson on the U.S. Annexation of parts of Mexico in 1848. Diplomatically stating that Dr. Goldberg must not have thought through his statement, Drolet said that he did not think Goldberg "really intends all the implications of such a statement." While pointing out errors in Goldberg's statistical analysis, Drolet thanked Goldberg for what he saw as the most important part of the study: the raw numbers of the Mexican survey. Drolet struck a balance between discrediting a study he saw as deeply flawed and maintaining a professional or scholarly tone.[29]

Goldberg's comments implied that Mexicans were racially and culturally inferior and that they migrated from the "uncivilized world." Mexicans were also racialized through the idea that they were, as a group, dirty. When applied to Mexicans, "dirty" also carried racial connotations of being darker-skinned and therefore inferior to whites. The refusal of the "boss" at the Cracker Jack Company in Chicago to hire darker-skinned Mexicans because he thought they looked dirty is one example; the attitude of the Polish boys in the Bush is another.[30] An employment manager at a large steel plant in Chicago was more explicit about his reasons for hiring lighter-skinned Mexicans. "It isn't that the lighter-colored ones are any better workers," he explained, but he had "noticed the attitude of our men when they ate in the company cafeteria" with African Americans and "chose Mexicans instead of Negroes." He hired only "lighter-colored" Mexicans "in order to minimize feelings of race friction and keep away from the color line as far as possible." The manager's attitude, shared by management at other industrial sites employing varying numbers of Mexicans, meant that skin color—even within this one ethnic group—mattered. Mexicans with darker skin encountered greater hostility and job discrimination than those with lighter skin.[31]

The racialization of Mexicans in Chicago became more pronounced over time as their numbers grew and the communities became more established. In 1928, an experienced settlement house worker observed: "Now the Mexicans are drawing off to themselves as there are more of them." She saw that "there is beginning to be race feeling" against Mexicans in Chicago; other immigrants "are beginning to say they are black." A group of Italians had come to her and said that if the University of Chicago Settlement House continued to rent their hall to Mexicans, they would no longer rent it for weddings. The worker told the Italian group that "In Italy you would not be prejudiced against the Mexicans because of their color." One replied, "No, but we are becoming Americanized."[32] As an increasing number of Mexicans entered the Chicago area, and pre-existing immigrant groups came to see them as a separate group, more and more southern and eastern European immigrants classified Mexicans as racially inferior. Not surprisingly, by doing so, newer southern and eastern European immigrants could move up the social ladder and closer to the white end of the racial spectrum.

Mexicans in South Chicago responded to this discrimination and oppression by means of traditional and non-traditional forms of resistance within the workplace and in other areas of daily life. Mexicans stood firm through individual workplace resistance, as well as through creating cultural, religious, social, and athletic organizations that served to unify and organize members of the Mexican community. They were also active in workplace organizing, but they did so largely through Mexican-only organizations and only to a much lesser extent through traditional labor unions. Understanding how individuals, small groups, and large groups practiced these forms of resistance is crucial to gaining insight in how Mexicans in South Chicago organized in order to survive.

Resistance was important not only in the everyday life and survival of Mexicans in the neighborhood of South Chicago. It was also a key factor in the ability of those in the community to persist and persevere. Anglo-Americans and ethnic Europeans perpetuated an environment of intolerance and inequity during both prosperous and bad economic times. Mexican resistance in South Chicago included actions that they might not have consciously considered acts of resistance. They fought small individual and group battles to combat harassment and reclaim their cultural, social, and economic terrain rooted in their individual identity and dignity. From quitting work when they objected to their conditions to outperforming ethnic European coworkers who harassed them, Mexicans actively resisted.

Scholars have a long tradition within the study of labor to examine informal acts of resistance in the industrial workplace. In the last three decades, scholars have more widely recognized that resistance does not have to consist of institutional, organized, conscious acts to be defined as resistance or to be effective. In the mid-1980s, anthropologist and political economist James C. Scott expanded the definition of what counted as resistance. He moved beyond the organized protests and institutional avenues more commonly recognized by historians to that point and argued that peasant resistance in differing locations encompassed "ordinary weapons" such as "foot-dragging, dissimulation, false compliance, pilfering, feigned ignorance, slander, arson, [and] sabotage." While subsequent scholars have criticized what they see as an overreliance on these everyday tactics as supplanting the influence of organized labor unions and other forms of pre-planned action, Scott himself was careful to state that his aim was to demonstrate that these "ordinary weapons" were a corollary to—not a replacement for—other, more highly organized forms of resistance.[33] Others have questioned whether the category of resistance remains useful when it includes such a wide range of actions.[34] Scott's research was largely on rural populations in Southeast Asia, but scholars of many specialties found his attention to everyday tactics insightful.[35]

Ethnic studies scholars and historians of minority populations in the United States have scrutinized and substantiated Scott's spectrum of resistance. Even though they do not all cite his work directly, they have corroborated the practice of broad-based, spontaneous, and often anonymous resistance of ethnic and racial minorities against groups and individuals with power over them.[36] While Vicki Ruiz examined union activity among Mexican women in the American Southwest, her analysis of women's "border journeys" that resisted dominant culture also included girls' embracing the flapper image while remaining under the eyes of attentive chaperones. In his study of Mexican citrus workers in greater Los Angeles, historian Matt García argues convincingly for making cultural acts of resistance a central part of any story about working-class racial minorities, so that what happened outside the union hall and workplace was integral rather than marginal to "collective expressions of resistance."[37]

Resistance did occur in the workplace, where Mexican workers sometimes cooperated to resist the harassment or discrimination by non-Mexican workers. Mexicans who were forced to work with ethnic European coworkers who harassed them and refused to help them "learn the ropes" resisted by turning to each other for support and encouragement. Although this can, at times, seem like nothing more than surviving and finding mutual support among friends, this was also an organized way to overcome harassment. In April 1923, when Benigno Castillo transferred from a U.S Steel job that paid regular hourly wages to one that was contract (piecemeal), he was upset that he "got paid less the first day because all the other *razas* didn't tell us how to start." In response, he and other inexperienced Mexican contract workers resolved to make more money than the other *razas.* "The next day," Castillo boasted, "they wanted to be friends. I said no," as did the other Mexican men.[38] This resistance permeated throughout public and private space.

Another resistance technique used by South Chicago Mexican workers was quitting *en masse* after a member of the group was mistreated or harassed. Like the previous example, this action can be considered nothing more than a group of people getting frustrated and doing what they saw was best for themselves and the group. Individually, these instances of quitting *en masse* do not look like acts of resistance. Collectively, as Mexicans gained a reputation of quitting as a group when they did not like how they were treated, foreman changed tactics for fear of losing their much needed workforce. Over time, Mexicans saw this action as an empowering form of resistance.

Gilbert Martínez gives an example of this when he witnessed a crew quit *en masse* in a performance of resistance against actions they deemed unfair. As highlighted in chapter five, after working for over a year and a half as the water boy for a Mexican track maintenance crew, Martínez witnessed the entire crew throw their shovels down and walk away. "That's the way it was then," Martínez explained. "The Mexicans didn't give a good god damn what [the foreman's] feelings were. Among the Mexicans, if they didn't like a guy they would find some way to, well, you know."[39] Despite liking his job and having a good relationship with the foreman, Martínez felt that he had no choice but to leave with the crew. This was another way *traqueros* made the workplace difficult for foreman who treated Mexicans poorly, even if they might have trouble finding another job. During the dark days of the Great Depression, South Chicago steelworker Benigno Castillo joined a group of Mexicans who quit Wisconsin Steel after two fellow Mexicans had been fired for talking. They did so knowing that finding a new job would be difficult.[40]

On the other hand, Mexicans also worked as a group to protect foremen they liked. After a well-liked "American" foreman in charge of a *traquero* work gang was fired because "a rail on our track was found by the roadmaster to be one inch from the end of another," the Mexicans refused to work. One of the workers explained, "The roadmaster tried to get the men out of the cars, but they asked for their checks before they would get out, so he gave us our time."[41] It is unclear whether the men were successful in getting their foreman back; however, they were successful in getting paid for the work they had done. These actions, stopping work in support of their foreman and refusing to leave their railroad cars until they were paid, demonstrated concerted resistance to what they saw as the unfair and arbitrary firing of a foreman who was friendly and fair to Mexican workers.

In addition to quitting as a group, Mexicans left individually. Experienced *traqueros* were less tolerant of abuse by foremen. A *traquero* interviewed by Paul Taylor in Chicago's Canal Street employment district discussed having left one crew because his foreman required extra work that went unpaid. He commented that he left his most recent job after only four days for the same reason. Adamant that he had learned his lesson the first time, he pronounced, "I should work for nothing! They can't get away with it with me any more!"[42] Although workers quitting every time they didn't like their working conditions might not be considered resistance, collective worker tactics that included work slowdowns or poor work performance that would reflect badly on the foreman are forms of resistance under Scott's theory.

As with other immigrant groups, Mexican-owned businesses could also become a locus of community for South Chicago Mexicans and a site of resistance to less direct cultural pressures. José Galindo opened a drugstore at 8901 Buffalo Ave in 1923, at the same address as his residence.[43] By 1926, "everyone from the neighborhood came to buy." Customers came more often to buy the herbs for familiar healing practices, rather than the medicines regulated by the American medical establishment. One customer who wished to purchase a medicine described the appearance of the container—"the medicine that is sold a little somewhat greenish bottle this size"—rather than asking for it by name, thereby indicating that even when people did use "American" medicines, they were still using them in their own cultural context. The store also drew Mexicans from as far away as Joliet, suggesting that there were not many options elsewhere that carried familiar products, was welcoming, and felt like a Mexican drugstore. One of the reasons people went to Galindo's drugstore, rather than the unfamiliar spaces that catered to the dominant culture, was that it was comfortable and "did not seem very elegant."[44] In addition to providing the materials with which to make remedies learned in Mexico or from relatives, the drugstore was a place South Chicago Mexicans could gather and converse with others, sharing news about friends, family, and work possibilities. The drugstore was a business, but it was also a shared community space within the neighborhood that offered Mexicans in the Chicago area the means to resist adopting one particular aspect of the dominant culture. By stocking different kinds of health remedies, the store enabled community members to avoid absorption into dominant ways of understanding health and illness.

Historian Lizabeth Cohen argues that the ethnic neighborhood store was an important part of many ethnic communities in Chicago. In *Making a*

New Deal, she establishes that going to a neighborhood store was not only about ethnic loyalty or the ability to talk to a merchant in one's native language. Immigrants shopped in stores operated by members of their own ethnic group because of trust. According to Cohen, residents of ethnically mixed neighborhoods sought out and patronized stores because they felt that they did not have to worry that merchants of their own group would try to cheat them.[45]

Mexican immigrants and Mexican Americans in Chicago created communities that brought their members together geographically. They also facilitated cultural solidarity in the face of what many Mexicans considered attacks on their heritage. Although external pressure to assimilate was substantial, Mexican men and women chose how, when, and to what extent to Americanize. However, with the community's ability to resist came expectations that members conform. Where the ability to resist existed, members of the community expected those around them to resist. Mexicans in South Chicago were expected to stand firm against Americanization efforts and not do anything neighbors and community leaders might construe as an attack on Mexican culture and community.

Reporting an example of transgression by members of the Mexican community in nearby Joliet, Illinois, a 1925 newspaper article in the Chicago paper *Mexico* warned of the shamefulness of turning one's back on one's *patria*, its customs, and its culture. After describing the "honorable" and humble Mexican who worked hard, celebrated being Mexican through his clothing and attention to Mexican customs and holidays, and sent money home to Mexico, the article contrasted the "bad" Mexican who left unsolved problems in Mexico, visited jazz clubs, and had forgotten the "little and bad" Spanish he learned in Mexico. The article's author then identified two of these "bad" Mexicans by name, Ambrosio Castíllo and José Saldivar, both interpreters for Illinois Steel Company in Joliet. In claiming to be "Spanish" instead of Mexican, and in distancing themselves from the Mexican workers at the plant, the article's author argues that these men had shamefully turned their backs on their *patria*. In addition to "calling out" these two men, the article further promoted the need for all "honorable" Mexicans to observe Mexican customs closely and avoid any behavior that might denigrate their Mexicanness.[46] Although the writer used *Mexicanos* in Joliet to make his point, his editorial was a call to Mexicans throughout the Chicago area.

The most prominent community organizing efforts focused on Mexicans proving their loyalty to each other and to their *patria*, rather than to their "white" neighbors and South Chicagoans at large. By forming a neighborhood-based Mexican community and alliances with other Mexican communities in the Chicago area, *Mexicanos* were able to create a culture that encouraged and supported mutual aid organizations, pro-Mexico patriotic clubs, social clubs, and athletic organizations. Within this Chicago-area Mexican "third space," communities were able to support newspapers that furthered the community's development and cohesion and allowed for public debate of key community issues. Outside attacks on Chicago's Mexican community and Mexican culture frequently united these heterogeneous and sometimes factionalized communities, even as organizations and newspapers came and went as a result of political infighting and economic interests. Historian Louise Año Nuevo-Kerr contends that Chicago's health, educational, and legal institutions frequently "threw up obstacles" that made Mexican socioeconomic progress difficult. *Mexicanos* responded to these obstacles through creating and actively participating in mutual aid societies, patriotic organizations, and through "religious affiliations."[47] During the Great Depression, Mexicans turned to these same organizations for social and economic help and in order to provide a more unified front against harassment and discrimination heightened by the economic crisis.

The importance of these organizations to Mexican cultural preservation and community formation, as well as the organizations' centrality in the daily lives of Mexicans in South Chicago, is apparent in the continuous newspaper coverage of the internal politics and operations of the organizations as well as their social, athletic, and cultural events. A September 1926 newspaper editorial in the *Correo Mexicano* advocated membership and attendance in group activities in the course of giving a brief history of several organizations. The writer emphasized the importance of the many organizations to the community: "To date, we have an infinite number of Mexican societies, clubs, athletic and cultural [organizations], and who knows what else, all of which are doing much to promote our patriotic culture, as much as humanly possible."[48]

Upset that most people outside of the community believed that the Mexican consul was responsible for the creation and promotion of many of these organization—an idea that the consul was slow to deny—the above editorial emphasized that members of the community were themselves responsible for the organizations and their activities.[49] By arguing community—not Mexican government—ownership of these organizations, the editorial emphasized the pride the community held in the fact that these organizations existed and were important to everyday life. That did not stop Mexicans from accepting consul support. While this editorial might have underlined the community's role in running the organizations, the leadership of those groups expected the consul to appear at their events, as well as to support their activities financially and by using its resources to promote events.

As was the case for other urban, ethnic immigrant communities, Chicago-area Mexicans formed social clubs, cultural societies, mutual aid societies, and other worker-based organizations for many reasons. *Mexicanos* formed and joined organizations to seek a better life, for economic or social security in case of illness, and for death benefits for their families. Mexican community organizations served as moral and cultural "flag-bearers" for the community, keeping watch over cultural expectations and activities. As the size of Chicago's Mexican community grew and its members became more entrenched—admittedly or not—in Mexican South Chicago, the number of support organizations also increased. Despite the fact that many Mexicans in South Chicago saw themselves as sojourners who would return to Mexico sooner rather than later, they established organizations to improve their quality of life while in South Chicago and maintain Mexican customs and a support base.

Between 1917 and 1928, Mexicans in Chicago started a total of thirty-nine organizations; twenty-six of them remained active in 1928. Of these, nine were in South Chicago, with only one listed as no longer active in 1928. The first Chicago Mexican organization was the *Sociedad Benito Juarez*, chartered in 1917, with the first South Chicago Mexican organization being the *Sociedad de Obreros Libres* (Free Workers Society), chartered in 1922. In addition to the above organizations, two Mexican Masons' lodges operated in Chicago, with the one in South Chicago dating back to 1920.[50]

Although much less publicized by the local Mexican newspapers, Catholic mutual aid and civic organizations were important components of life for many *Mexicanos* in South Chicago. The neighborhood was, as mentioned earlier, home to Our Lady of Guadalupe, the first Chicago Catholic church built specifically for the Mexican community, and by 1928, it had become a large, popular symbol of Mexican Catholicism in South Chicago. Organizations based at Our Lady of Guadalupe refused to participate in a short-lived confederation of Mexican societies organized by the Mexican consul in 1925. The Claretians and their churches had little contact with the Mexican Consul because of the order's history of being expelled from Mexico and the Mexican government's continued anti-Catholic rhetoric and actions. This strife between the Catholic Church and the Mexican government was an important reason that local newspapers, primarily the liberal *Mexico*, ignored parish news and organizations.[51]

Catholics created two parish-based organizations soon after Father Kane founded Our Lady of Guadalupe: *Los Caballeros de Guadalupe* and *Las Hijas de Maria.* Like the Protestant organizations discussed earlier, *Los Caballeros de Guadalupe* sponsored neighborhood dances and celebrations. Members of the church also established a church-sponsored mutual aid society, but that organization split from the church when Father James Tort insisted on control of the organization.[52]

Another early Catholic South Chicago organization was *La Union Nacionalista Mexicana.* Although it is not clear when the organization was started, it was in existence in 1920. Its mission, according to Robert C. Jones, was "to help bring about the downfall of the present liberal government in Mexico." This organization was yet another example of the close political and social ties between Mexican Chicagoans and Mexico and their continued desire to keep a hand in Mexican domestic affairs.

Mexicans created these organizations, social clubs, mutual aid societies, *pro-patria* associations, and community service groups, not only because of their large numbers, but to combat the environmental, social, and economic isolation caused by white and ethnic European racism and discrimination. *Mexicanos* also felt a need to promote Mexican culture while serving those in the community. Mexicans equated discrimination and most assimilation efforts as not only an attack on them and their skin color but on their culture. In response, they created organizations that celebrated Mexican culture and Mexican national holidays, a goal that the Mexican consul in Chicago actively encouraged through his vocal presence at events and his encouragement of the formation of organizations.

Despite a desire to show autonomy from the Mexican government, local leaders expected the counsel to participate, promote, and co-sponsor society events. Meeting local expectations, the consul actively encouraged participation in mutual aid societies. At a celebration in honor of the September 16th holiday held in 1924 sponsored by the *Mutualista Benito Juarez* in "cooperation" with Mexican consul Luis Lupián, the consul played a prominent role in promoting the organization's work among the 1,000 Mexican attendees.[53] Interviews, newspaper articles, and archival sources show the consul's continual active support of community organizations.[54]

This support, though, came with the expectation that he would go beyond simply promoting and financially backing an event. *Mexicanos* expected the Mexican consul to actively represent the Mexican state at community meetings. In one example, the failure of the consul to appear at a city-wide meeting of all local chapters of the *Frente Popular* led to demands for his recall to Mexico City. At the meeting held at Hull House on February 2, 1936, a group leader made a "fiery and impassioned speech" criticizing the consul for not attending. The consul, having accepted the invitation to appear, failed to show, and "as this was not the first time he had disappointed them a motion was made and carried that a letter be sent him demanding an explanation of his conduct and insisting that it was the right of the Colony to demand his presence among them when they thought it necessary." The group agreed that if the consul was unwilling to appear at a meeting, the group would "ask Mexico City of his recall."[55] The consul apologized for his absence, and the matter was later resolved to the satisfaction of the *Frente Popular* leadership. These organizations assisted in the creation of a community that was able to resist the wholesale efforts at Americanization, and they expected the local representative of the Mexican government to support their activities.

Mexicans in Chicago formed *mutualistas* to help provide economic security and as a platform for cultural and social resistance in support of members of the community. For a minimal monthly fee, members were, in theory, assured of aid if unable to work or in case of death for any reason. From the early years of the Mexican community of the Chicago area, mutual aid societies emerged as organizations that attempted to improve the lives of those in the community through protection of workers. The *Sociedad Mutualista Benito Júarez* was established in 1917 by a group of Mexican "workers, lawyers, doctors, engineers, businessmen, writers, and newspapermen."[56] This *mutualista* served as a longstanding mutual aid society in addition to a social organization that provided members a forum for discussion and sponsorship of community activities. Another such organization, founded in the early 1920s, was the *Sociedad Fraternal Mexicana*. It too served as a social organization in addition to a mutual aid society.[57] It is important to note that *mutualistas* frequently served as launching pads for community activism in support of the Mexican community. Two examples of South Chicago-based mutual aid societies that held public community social events in addition to their mutual aid functions were the *Mutualista Plutarco Elias Calles* and Camp Emiliano Carranza of the Woodmen of the World.[58]

Another longstanding *mutualista*, the *Obreros Libres Mexicanos* (Free Mexican Workers), was particularly active within South Chicago's Mexican community. In 1928 and 1934, leaders posted announcements in local Spanish-language newspapers urging workers to join the organization. They asked people to become a part of the work that was being performed by the society and urged them to enroll under "the *mutualista* flag" that was "well known to state authorities."[59] The *Obreros Libres* sponsored neighborhood and city-wide *Cinco de Mayo* celebrations as well as sponsoring its own *orquesta tipica*. The creation in 1929 of the *Orquesta Tipica Obreros Libres Mexicanos de South Chicago*, a string band that played popular and patriotic music, was right in line with the organization's stated purpose of improving the Mexican community through culture and economic aid.[60]

In a notice published in a November 1929 issue of the local newspaper *Mexico,* bandleader Narciso González of 8242 Mackinaw Avenue espoused that community support of the *orquesta* was an important and patriotic duty. Explaining that it was "the inalienable duty of every good Mexican to contribute some little effort to the prestige of our dear and beautiful country," González asked for the Mexican community's "personal contribution or its moral and material aid . . . so that with fraternal harmony we may form a homogeneous group that may accomplish our final end, that is, to give our best efforts for the good of our beloved country." As representatives of the community, the *Obreros Libres* embodied the priorities that many *Mexicanos* believed important: unification and improvement of the Chicago Mexican community and the promotion of the parts of Mexican culture they felt most important. They linked all of this to patriotic loyalty and the immigrants' sojourner attitude, making these actions a duty and responsibility of every *true* Mexican.[61]

Members of mutual aid societies took advantage of the organizations' ability to bring workers and their families together to make *mutualistas* much more than insurance cooperatives. The sponsorship by the *Mutualista Ignacio Zaragoza* of José Vasconcelos's speech on the priorities and perseverance of Mexicans in Chicago exemplified the broader outreach activities of such societies. It was not uncommon for *mutualistas* to sponsor or hold pro-Mexico events, holiday celebrations, or talks. They could also serve as a united voice for members on issues important to the community. Although the *mutualista* Benito Júarez organized English classes for its members, not all thought that such classes would be helpful. Carlos Perez Lopez, suit salesman and president of the *Sociedad Fraternal Mexicana* in Chicago, observed that the majority of workers in unskilled positions "are neither able to nor pretend to learn English."[62] Perez Lopez's comments highlight the class divisions within the community and the simple fact that the greater Mexican community in Chicago was heterogeneous.

To keep ties between Mexicans in Chicago and in Mexico strong, the Mexican consul acted as protector of the community and continued to encourage people's return to their home states. In addition to visible participation in community events, the consul used local Spanish-language newspapers to remind Mexicans in Chicago, including those in South Chicago, of the importance of staying connected with the *patria.* Newspaper editorials reminded Mexicans to register at the Mexican consulate and "get a certificate of citizenship" to make traveling into Mexico easier. Editorial writers also warned that lack or registration would, in the course of time, lead to the loss of Mexican citizenship.[63] Although the consul pushed for the maintenance of Mexican cultural and political ties, he did not discount the importance of learning English in the survival of the Mexican community of Chicago.

Despite the strong network of organizations, community observers and advocates—as well as some Spanish-language newpaper editorials—criticized the large number of small societies as a sign of poor intra-community cooperation. They argued that too many *mutualistas* meant that most of them remained small. Small mutual aid societies were at risk of being unable to collect sufficient dues to be financially sound enough to pay out more than a few benefit checks before having to fold.

Interviewed around 1930, a Mexican immigrant identified as I. Ramos complained that one of the first things he noticed when moving to Chicago was that his *paisanos* "were divided into little groups. One group was interested in this society, and the other in the other." The ability to choose among patriotic societies, mutual aid societies, and religious organizations was new to Mexicans, he argued. "There had never been an opportunity to take part in such societies out in the country districts where I came from and where the majority of my countrymen here come from. There may have been societies in the towns but we never heard of them. There weren't even any societies connected with the church." These new opportunities—too many, according to Ramos—were used by many former *campesinos* to adapt, survive, and cope in their new urban settings.[64]

In 1928, a Claretian priest at Saint Francis of Assisi Church, the Mexican Catholic church in Chicago's Near West Side neighborhood, complained that the city-wide Mexican community lacked cohesion and could not control political infighting. Father Domingo Zaldivares, a Spaniard, argued that only religious societies did well in Mexican communities. At first, he "was quite enthused over their mutualistic societies and their efforts to help one another" but he quickly changed his mind as he saw the self-imposed limits and infighting within these mutual aid societies. He criticized the societies for limiting their membership to Mexicans and creating several mutual aid socities instead of one bigger, more financially viable entity. Zaldivares made the same generalizations as many in the dominant community who tended to view all Mexicans in Chicago as one homogenous group. He complained that the "Latin race has that tendency to be so divided and split up and not capable of any great united efforts like the other peoples undertake and do."[65] From his perspective as a Catholic priest, he believed the Catholic Church and Catholic-parishioner societies could satisfy unmet needs and therefore Mexicans should turn to the church.

Many immigrant advocates, social workers, and some members of the Mexican community echoed Zaldivares' complaint that too many small mutual aid societies forced the financial collapse of most of them.[66] The 1928 president of the *Circulo Azteca*, a South Chicago social and cultural society, gave a formal address on the dissension within the Mexican community. At a city-wide gathering of Mexican associations held at Hull House's Bowen Hall on the Near West Side, he asked those present to be prepared and alert "to oust from our midst those who would plant and sow the discord, the ill-will, the break-up of our friendships and our purposes." Worried about the perception of the Mexican community among the "Americans," he argued that there were "many who would join us, agree with us, only later to take hold of the reins and by disgraceful acts discredit us not only before our own people but amongst Americans." Continuing, he argued that a disunited Mexican community could only benefit the Americans. "There are powerful interests here in Chicago and elsewhere who look with jealous eyes on our efforts to unite and better ourselves. It is more than often that other emmisaries come amongst us to disorganize and disrupt us. When they do come, let us pick them out, and step on them hard and cast them out."[67] The speaker sought to unify the Mexicans in Chicago by rallying community members to act against a covert attack by those who would harm them.

Mutualistas in Chicago, large and small, occasionally overcame political and social differences to unite. These organizations served assorted roles for their dues-paying members and the wider Mexican community. In addition to their function as insurance providers, *mutualistas* sponsored social gatherings that provided venues for social exchange and community building. They trumpeted loyalty to the *patria* as the reason to come together and sponsor cultural celebrations.

Mexican steelworkers in South Chicago turned to *mutualistas* for some of the same reasons other workers joined labor unions. Widespread labor union discrimination and harassment against Mexican workers was one reason *Mexicano* steelworkers turned to the Mexican-only societies for support and to organize against workplace grievances. Although the vast majority of Mexican steelworkers would eventually become members of the Steelworker Organizing Committee (SWOC), recruitment of and participation by Mexicans was limited before 1936.[68] Discrimination experienced by rank-and-file Mexican workers—as well as Mexican workers who helped in the organizing drives—made organized labor a limited, albeit necessary, form of resistance.

In 1928 there were few Chicago *Mexicanos* involved in trade unions since many worked in the poorly organized steel and meatpacking industries.

Even when unions did exist in a particular industry, they were often weak in the low and unskilled grades in which most Mexicans were occupied. Thus, very few track laborers were part of the Brotherhood of Maintenance of Way Employees. Other unions, such as those for the building trades, required American citizenship for membership, thereby preventing most Mexicans from joining even had they wanted to.[69]

Employers recognized that *mutualistas* filled many of the same functions for Mexicans as unions did for other workers. A labor agent for a sugar beet company that recruited in the Chicago-Gary region claimed to see dangers of communism in any type of organization. He said, "I am absolutely opposed to the Mexicans organizing, regardless of the object. These organizations begin innocently with literary purposes but they soon developed along Bolshevistic lines of thought." An employment manager for a company with a large proportion of Mexican employees made a similar point about *mutualistas* leading to other kinds of organizing. He said, "The Mexicans have societies, and of course if they organize one way, it is but a step to organize another way."[70] These were strong words implying that Mexicans could not be trusted during this red scare. Any social organizing among Mexicans, this manager argued, would easily lead to the development of political, anticapitalist organizations.

In some cases, Mexican workers sought to remedy difficulties left unaddressed—or even caused by—union organizers and members. Although the majority of Mexican steelworkers in South Chicago joined the SWOC, many were frustrated by the widespread prejudice and unfairness. As a result, some who had started off as active organizers chose to limit their time spent doing union work. In the fall of 1935, Alfredo De Avila, one of the early Mexican organizers for the SWOC, started as one of about twenty-five to thirty South Chicago Mexicans who volunteered to organize during the early stages of the SWOC campaign.[71] De Avila, soon after starting his involvement with the SWOC, accepted a position as a full-time paid organizer with the union. After organizing seven days a week for six or seven months, De Avila asked to be a part-time organizer and return to the plant because he "saw the anti-Mexican discrimination didn't cease. Even in the union, I feared I wouldn't last long there."[72] Although De Avila was a successful organizer and believed in the union and its principles, he decided to return to work at the steel mill and lessen his time commitment to the union because of constant harassment by some non-Mexican labor organizers and SWOC officers. His long-time activism for the SWOC and his short stint as a full-time organizer were enough to lead to a federal investigation to find out if he was a communist when he applied for U.S. citizenship in the early 1940s. He was cleared and

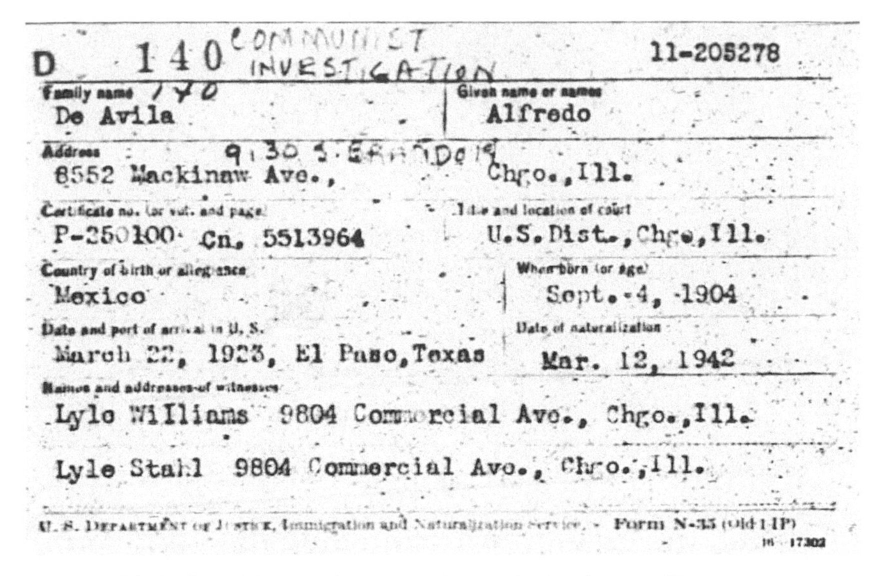
D 140 COMMUNIST INVESTIGATION 11-205278

Family name 170
De Avila

Given name or names
Alfredo

Address 9130 S. [illegible]
8552 Mackinaw Ave., Chgo., Ill.

Certificate no. (or vol. and page)
P-250100 Cn. 5513964

Title and location of court
U.S. Dist., Chgo., Ill.

Country of birth or allegiance
Mexico

When born (or age)
Sept. 4, 1904

Date and port of arrival in U. S.
March 22, 1923, El Paso, Texas

Date of naturalization
Mar. 12, 1942

Names and addresses of witnesses
Lyle Williams 9804 Commercial Ave., Chgo., Ill.
Lyle Stahl 9804 Commercial Ave., Chgo., Ill.

U. S. Department of Justice, Immigration and Naturalization Service. Form N-35 (Old 1-IP)
16—17302

Fig. 6.1. Alfredo de Avila's naturalization card. Note the handwritten "Communist Investigation" note. It was not uncommon for union activists to be suspected communists. His naturalization is during World War II, a time when many Mexicans in the steel industry were pressured to become citizens. Soundex Index to Naturalization Petitions for the United States District and Circuit Courts, Northern District of Illinois and Immigration and Naturalization Service District 9, 1840–1950. National Archives and Records Administration, Washington, DC. Microfilm: M1285 Roll 40.

granted citizenship in 1942. Despite and because of harassment by coworkers, union organizers, company managers, and the government, Mexicans continued to join the union. They also continued to organize and resist through Mexican-led and Mexican-run organizations before, during, and after the SWOC campaigns.[73]

Cultural celebrations also provided Mexicans and their *mutualistas* a way to resist the ever-present harassment and to aid in the creation of Mexican-friendly physical and cultural environments. Mexican cultural ideals brought to and developed in South Chicago were yet another iteration of "Mexican culture" and distinct from other Mexican communities in the United States. Members of the community embraced Mexican cultural celebrations and practices promoted by the *mutualistas,* the Mexican consul, and other local patriotic organizations. This pushed Mexicans to look inward in defense of their cultural practices and celebrations as functions of their fundamental ideals in the face of a constant pressure to Americanize. In assimilating only to the extent necessary to survive and improve their quality of life, Mexican

Chicagoans created an idealized Mexican culture distinct to Chicago, one reinforced by the Mexican state through the local consul.

On the other hand, too much public celebrating had drawbacks regardless of the "third-space" celebrations occupied. Worried that Mexicans would be seen as bad, lazy workers, the *Correo Mexicano* newspaper urged its readers to go back to work immediately after the September 16th Mexican independence day celebrations in 1926. After members of the community had "complied with one of the most sacred duties of being Mexican" by honoring "the men of the *patria,* the heroes that offered their lives so that we may have ours," it was time to get back to work, "to follow the trail blazed by the books, by the moralists and by the teachers of the secondary schools and universities."[74] On top of revealing the fears of middle-class Mexicans about how outsiders would perceive the community if members of the working-class took too much time to celebrate, this editorial is significant because it provides a glimpse of the scale in which members of the community participated in the celebrations. More than a light-hearted fun time, participation in the celebrations was a "sacred" duty. The *Correo* thus espoused the tenets of a civic religion centered on the Mexican state and those who had created it. First among those honored were, of course, the warrior "heroes" who had died for the *patria* and thus enabled the immigrants' lives in 1926. Mexicans' next responsibility, however, was to honor the work of writers, philosophers, teachers, and professors who had defined the path to the creation of a nation. The editorial drove home the point that the *Correo*'s readers might best continue these intellectuals' legacy by following their trail back to work—that is, to labor that was not so incidentally quite far afield from those who "blazed" the trail.

Linking the *Diez-y-Seis de Septiembre* holiday to the need to learn English was important because of the hallowed space this holiday held in the psyche of Mexicans in Chicago. *Diez-y-Seis de Septiembre* was honored as the highest Mexican national holiday and was actively observed by most local Mexicans. *Mexicanos* from South Chicago attended large, central celebrations as well as smaller, community-based ones held in neighborhood parks or dance halls. The large, central celebrations could include as many as 1,000 participants at the 1924 celebration held in the West End Women's Club and 5,000 participants held in Pilsen Park that same year. Most of these celebrations usually included a band or orchestra, singers, patriotic readings, and speeches by local Mexican community leaders. By the middle to late 1920s, the central celebrations, first hosted by the *Mutualista Benito Juarez,* were organized by a commission comprised of leaders of Mexican organizations from all three Chicago Mexican communities and always included the official presence of the Mexican consul. Smaller, neighborhood-based celebrations usually occurred on the same evening. In 1927, for example, South Chicago Mexican youth organized a neighborhood celebration that included the *orquesta Azteca* and many of the same elements of the larger celebration including the singing, patriotic readings, and speeches.[75]

These celebrations, primarily for the two high Mexican patriotic holidays of September 16th and May 5th, became a way to culturally resist assimilation and to resist a diminishment of Mexican cultural identity by celebrating these holidays in a very public way. The celebrations helped to create a sense of national identity among participants, one rooted in Chicago that at least nominally reached across regional and racial differences among Mexicans. The Mexican Revolution and its aftermath resulted in a cultural flux and instability that further factored into the variations of interpretation of Mexican culture in U.S. cities.

Historian George Sánchez contends that Mexicans in Los Angeles invented new traditions and discarded or radically transformed older customs at the same time that Mexicans in their home country were "creating 'traditions' to cement national identity." This was also true in South Chicago. As discussed earlier, Sánchez called this zone where that from "Mexico"

Fig. 6.2. Lupe Vera's Orchestra, South Chicago neighborhood group; piano player is Tino Cordero, son of Justino and Caroline Cordero. Photograph courtesy of the Southeast Chicago Historical Society (neg. no 81-120-41).

met that from *Mexico de afuera* a "third-space" that was distinct to the local immigrant community. Although the idea of amalgamated and romanticized traditions is not unique to the Mexican immigrant community, Sánchez argued that Mexicans outside of Mexico were more predisposed to this practice than those from many other nations because Mexico "was a national community that had to be 'imagined' to exist, particularly given its racial and regional diversity."[76] Because people brought with them the definitions of "Mexico" and "Mexican" current in Mexico at the time of their departure, the constant pre—Great Depression migration of Mexicans to Chicago contributed to the continuing fluidity of evolving cultural traditions in South Chicago. In other words, Mexican South Chicago's "third-space" was constantly changing as the local culture adapted to external pressures and absorbed new migrants into the community.

Celebrations and resistance to outside pressures did tend to bring the heterogeneous Mexican community together, but Mexicans in Chicago frequently clashed when they looked inward. Class and regional tensions often divided the community. From time to time, elite *Mexicanos* echoed the negative stereotypes of lower- and working-class Mexicans promulgated by the dominant societies of the United States and Mexico.[77] Agustin Fink's discussion of his clients' approach to life insurance demonstrates the prejudice that whites and elite Mexicans directed at working-class Mexicans. According to Fink, Mexicans tended to take out short-term policies of five or ten years rather than long-term life insurance. Initially, the insurance company did not want to cover Mexicans because it "worried that their health conditions would not make them insurable." Because of this, Fink's first three clients were rejected. However, once the company doctor gave favorable reports on the Mexicans' condition, there was "no longer any resistance from the company about accepting them." Assumptions about Mexicans' poor health made it difficult to get insurance even for those few who chose to put some of their money towards insurance. These prejudiced assumptions were not only coming from Anglo Americans who controlled Aetna, or from others.

Class differences within the Mexican community in Chicago meant that more highly educated and elite Mexicans often looked down on the working class and poor. Agustin Fink, whose father was German, had been educated as a civil engineer at a German academy in Mexico before working as a vice consul of Argentina in Chicago. He described himself as "having nationalist tendencies," but also believed that "the American influence on immigrants contributes to their progress." In his view, only those Mexicans who had been in the United States for a while "understood the advantages of becoming insured"; those who had been in the country for only a short while could not plan for the long term. He himself arrived in 1923 and by 1930 had already filed "first papers" to initiate the naturalization process.[78]

One community leader, Carlos Perez Lopez, argued that "humble" Mexicans never denied their Mexican nationality, but "those of higher class and those who can pass [because of lighter skin] as Spaniards call themselves Spanish."[79] Echoing the sentiments of the *Correo Mexicano* editorial that passing as Spanish was bad, Perez Lopez implied that since the working-class, darker-skinned Mexicans did not have the option of passing as Spaniards, their loyalty to their *patria* remained more constant. Perez Lopez oversimplified the complicated dilemma many Chicago Mexicans faced. Some of those who passed as Spaniards outside of the Mexican community might still have felt allegiance to Mexico; for these individuals, claiming Spanish identity was a way to gain access to social and workplace opportunities that were not open to Mexicans.[80] Since most middle- and upper-class Mexicans tended to be lighter skinned, the ability of those lighter-skinned to pass as European led to class animosity and aggravated divisions within Mexican Chicago.

The lasting effects of the Mexican Revolution further complicated class relations within the Mexican community. A significant percentage of the Mexican community in Chicago was composed of those who—in the words of dentist Juan Medina—had held a "certain social position" and then lost it, either directly because of their political affiliations before the revolution or more generally in the economic turmoil. Forced into a "more humble position," these Chicago Mexicans tended not to mingle with other Mexicans out of pride, nor even to ask for help. Medina estimated that as much as 15 percent of the Mexican population in Chicago belonged to this group.[81]

Regional differences were yet another major source of tension within the Mexican community. In his 1980 interview with Jesse Escalante, Max Guzman remembered the breakdown of South Chicago into concentrations based on home state and even specific cities of origin. He explained that "in 1927 or 1928," Buffalo Avenue marked the dividing line between a group from the town of Techaluta, Jalisco, and those from Zacatecas or Michoacán. Guzman's coworkers at Interstate Steel also believed that Mexicans continued regional rivalries in Chicago. Guzman claimed Mexicans "were one of the worst" ethnic groups in terms of not being able to get along with those from other areas of their home country.[82] Gilbert Martínez recalled a similar breakdown among South Chicago Mexicans. "At that time, there were cliques of Mexican *paisanos*. There were cliques of [those from] Jalisco, of *Durangeños*, and cliques of the *Norteños*." The state of Durango is located in the northwestern part of Mexico, just south of the border states considered home to *Norteños* and north and slightly west of the central plateau state of Jalisco. Those from Jalisco, particularly "*los Techalutas*," occupied "mostly all Green Bay [Avenue] and Mackinaw [Avenue]." Martínez, along with others interviewed by Jesse Escalante, characterized those from Techaluta as "rowdy and racket noisemakers." After payday, *los Techalutas* claimed a pool hall on Green Bay Avenue as their own; as Martínez put it, "you could go in there, but you went at your own risk."[83]

Mexicans in South Chicago—spurred by the reinforcing stimulus and revolutionary ideals provided by the Mexican consul and speakers such as José Vasconcelos—created sites of resistance that solidified their community. They resisted external pressures to "Americanize" whenever possible and refused, in any significant numbers, to seek naturalization. Many Mexicans in South Chicago believed it was their duty to fight to keep a strong Mexican cultural identity within the community. Even as some of these cultural practices were oppressive and discriminatory to segments of the community, most South Chicago working-class Mexicans understood—or at least believed in—resistance and organization in order to improve their everyday lives in South Chicago while maintaining their Mexican identity.

In going beyond the question of whether organizations such as labor unions, community-based groups or mutual aid societies spoke for the majority of Mexicans in South Chicago, it is important to look at these organizations as pieces of a puzzle where the most dynamic struggles took place outside of, or in spite of, the established traditional organizations. By recognizing and not underestimating the significance of everyday forms of resistance and the politics of culture, as well as institutions and organizations not normally seen as vehicles for everyday and working-class change, we can delve into the strategies that helped Mexicans in South Chicago cope with the oppressive environment that surrounded them. Labor union hostility toward Mexican workers further reinforced the importance of alternative sites of resistance. These sites included *mutualistas*, *pro-patria* organizations and local Spanish-language newspapers.

III

ENDURANCE

7
The Great Depression

Fifty years after the Great Depression, José Cruz Díaz still remembered Mercedes Rios and her perseverance in helping *Mexicanos* in South Chicago find work and navigate the relief bureaucracy while the Mexican consul focused on securing funding for a voluntary repatriation program.[1] Rios, born in San Antonio, Texas, first got involved in helping *Mexicanos* get relief in November 1932, when she stopped by the local public aid office at the request of her mother. She went to help interpret between aid workers and Spanish speakers seeking relief. Rios volunteered despite already working three days a week and expecting to return soon to a full-time work schedule. At the local public aid office, Rios found two lines: one for Spanish speakers and the other for everyone else. Spanish speakers faced longer wait times or were told to go home because interpreters were not regularly available.[2]

Upon arriving at the aid office, Rios was immediately put to work as one of only a handful of English/Spanish bilingual aid workers or volunteers in the city. She performed an essential role in assisting non-English speaking Mexicans to apply for aid. After spending three weeks as a volunteer interpreter, Rios returned to a six-day workweek at her speaker-factory job. Rios's supervisor at the aid office offered her a paid position in an attempt to keep her. Despite a pay cut, Rios agreed to work at the aid office, where her supervisor expanded her duties. She proudly recalled her work: "I learned so much that they sent me to the homes. They gave me a family. I went to the dime store to buy a little black notebook for 25 cents. And then I went from home to home." Buying the little black notebook was an important step for Rios and her role as a community leader as it signified the unofficial badge of a caseworker. She gained the independence and trust from her bosses and from the community.[3]

Many Mexicans in South Chicago considered Mercedes Rios a community leader. She was an English-speaking *Tejana* member of a working-class Mexican community who made sacrifices and advocated for other working-class Mexicans. Her education, her Mexican-American status, her family's historical and economic background, and her ability to interact with those outside of the community made her part of a small middle-class within a working-class Mexican South Chicago. Although Rios might have argued against her separation from the working classes, Rios's white-collar skills and leadership set her apart. Because of her leadership skills, Mexicans and those outside of the community recognized her as an authority figure and someone closely tied to the community. She could help those in need negotiate and navigate the government and workplace bureaucracies.

As an aid worker and community leader, Rios developed relationships with neighborhood shop owners and service providers. For example, she sought out the owner of a local coal yard: "I went over there and got acquainted with him and told him, 'When I call you, give them coal. I work for the office, and you'll get the disbursing orders.'" She also instructed Ruben Flores of the Tienda Colorado to "just take care of them and just call me and let me know and give me the receipt and then we will make the disbursing orders." Rios also worked closely with shop-owner and landlord Helena Svalina, a friend of the community not only in the early days of settlement, but also throughout the Great Depression. Rios and others lauded Svalina as the first non-Mexican merchant in the area to grant credit to Mexicans while selling Mexican food staples.[4]

Feeling that she had to do more than just help those in the community fill out aid paperwork and secure essential commodities, Rios built a reputation within Mexican South Chicago as the person who could get people jobs. She started by finding work for her brothers, which, in turn, enabled them to contribute to the household income. "I got acquainted with the superintendents at the [steel] mills, Mr. Bradbury and Mr. Conrad. I got my brothers jobs there when they were old enough to work, then I helped other Mexicans to get jobs there."[5] While it is unclear how she was able to gain access to and develop an influence with the mill superintendents, her language ability and her persistence undoubtedly made superintendents more willing to lend her an ear and then to keep listening. Once she gained access to the superintendents, Rios' concern went beyond just securing any job for *Mexicanos*. Upset that the majority of Mexicans were being hired only for lower-paying outside jobs, such as track worker, Rios protested to the mill superintendent: "I said 'Mr. Bradbury, my people can do work inside the mill as well as the others.' So they started hiring."[6] Rios's status as an English-speaking member of the community, her position as a social worker, and her persistence, afforded her rare relations with mill managers. This was something she was acutely aware of as she worked to improve the lives of those in the community.

Examining individual Great Depression–era experiences is important to better understand how community functioned, how Mexicans were able survive in Chicago, and how they affected their environment. Mercedes Rios came of age during the depression. Her experiences are significant because she faced many of the same challenges as others in Mexican South Chicago and a more complete record exists of her life than that of the vast majority of those who lived through the period. José Cruz Díaz remembered Mercedes Rios and her perseverance in finding jobs during the depression for members of the Mexican community in South Chicago. Díaz recounted that after having met Rios, "I asked her one day to get me a job, and sure enough she got me a job. I didn't pass the first time I went, I was too skinny. But she put the pressure on and I got the job."[7] The story of this young woman's life and her struggle to improve the lives of people in her community during the depression has implications beyond that single life.

In many ways, Mercedes Rios's life exemplifies the history of the community. She and her family arrived in South Chicago during the early years of settlement, she came of age during the depression, and she emerged as an established leader able to negotiate for members of the community. Although she was only one person who made particular choices, in her story we can see many of the issues that faced the entire community.

Despite the comparative lack of involuntary repatriation programs in South Chicago during the Great Depression, the economic crisis had a deep and lasting impact. It shaped the Mexican community into one much different than before the depression. Both this chapter and the one that follows focus on events in the Mexican community of South Chicago during the period immediately leading up to and including the Great Depression. These two chapters explore how Mexicans in and around South Chicago continuously shaped their everyday lives to adapt to the evolving realities of economic turbulence in an unpredictable and often hostile environment by doing whatever was necessary to survive. For some Mexicans in South Chicago, this meant that multiple members of the same family, men and women, actively sought, and sometimes gained, employment. For others, it meant ignoring culturally reinforced norms to go on relief. For many it meant leaving the industrial heartland of the United States for Mexico, either willingly or unwillingly. The focus of the next chapter is organized sports and leisure among Mexicans in South Chicago during the Great Depression; in these arenas, men were able to claim rights as residents of the United States in ways they were unable to at work and in the neighborhood.

In this chapter, I argue that the events of the Great Depression in the Chicago area, including unemployment and repatriation, were critical in the evolution of a strong Mexican community of South Chicago that was smaller, had clear internal leadership, and was more physically and culturally ensconced in the neighborhood. Although involuntary repatriation programs were not as prevalent in Chicago compared to other cities with large Mexican communities, depression-era unemployment affected Mexicans in South Chicago to at least the same degree that it affected Mexicans in other parts of the United States. As unemployment rose, many Mexicans migrated out of South Chicago to other neighborhoods, to settle with family or friends in other parts of the United States, and to return to Mexico.[8] In this chapter, I also examine the reasons and realities that kept over 13,000 Mexicans in Chicago despite the economic turmoil and repatriation pressures.

Understanding the various forms of repatriation is important to understanding the complexity of Mexican repatriation during the Great Depression. Despite the predominant sojourner attitude in the Mexican community, people wanted to stay in Chicago until *they* were ready to leave. Many people were given the opportunity to leave during the depression; others were not given a choice. The three basic forms of repatriation were forced, coerced, and voluntary. Forced repatriation is the simplest form to understand. When someone forces a *Mexicano* onto a train, or any other form of transportation to the border, the *Mexicano* is being forced to repatriate to Mexico. In some places, this happened through government round-ups; in other places, local organizations, like American Legion posts, took charge of forced repatriation campaigns. Labeling someone's repatriation as coerced is a bit more subjective. Mexicans and Mexican Americans were fired, harassed, denied benefits and aid they were eligible for—and were submitted to other pressures—to convince them to return to Mexico. At times the coercion would be linked to free or reduced train fare to make their decision easier. The third form of repatriation, and the one Chicago's Mexican repatriates experienced most frequently, is voluntary repatriation. This form implies that there was no formal coercion to leave, but Mexican immigrants took advantage of free or discounted transportation and incentives to return to Mexico. Many of those who left Chicago decided that subsistence living in rural Mexico was much easier than unemployed life in Chicago. They might not have left on the terms they had originally planned to leave on, but they changed the plan and voluntarily left.

For Mexicans in South Chicago, organized and individual resistance to harassment and discrimination before the Great Depression—and the community's ability to coordinate for socio-economic support and cultural preservation—meant that community members had already created an infrastructure and had experience that helped leaders organize the community and come together quickly during the crisis of the 1930s. The need for leaders and organizers from within the local Mexican community was most pronounced in times of economic crisis, when those in the greater "American" society often blamed immigrants for taking jobs from white Americans and lowering wages because of their presence (or perceived presence) in the workplace.

Max Guzman's experience was not uncommon. He faced and witnessed the constant harassment of Mexicans in South Chicago during the Great Depression. Guzman, in his Escalante interview, argued that although forced repatriation was not the norm in South Chicago, community members lived in an environment where "a lot of Mexican people were sent back to Mexico" and where the authorities "were really cracking down." Guzman remembered that many Mexicans wanted to return to Mexico: "a lot [of Mexicans] were asked if they wanted to go back and a lot of them wanted to go back." Although he was never forcibly repatriated, Guzman has vivid memories of having to be in a constantly high state of alertness against local police and immigration agents. He was acutely aware of the possibility of being stopped and harassed anywhere at any time: "There was a car, I think it was called number four. That car used to be all over, they would stop you every place. They would take you over to court and they wouldn't hear nothing, Judge Green would say 'you, you, you,' and that would be it."[9]

Municipal Court Judge Thomas A. Green's aura of power and authority was so pervasive that his name and actions continue to be part of Mexican South Chicago folklore. Several South Chicago Mexicans interviewed five decades after his time on the bench remembered Judge Green and his maltreatment of Mexicans in the community. Indeed, his judicial arrogance extended to jailing the acting Mexican consul, Adolfo Dominguez, when the diplomat protested against Green's insulting Mexico and Mexicans.[10]

On July 7, 1931, Consul Dominguez attended a hearing in Judge Green's court for an unemployed Mexican immigrant who had been arrested in South Chicago for vagrancy. After arraigning five Mexican "idlers," the sixth person complained he had no job, home, or friends in Chicago. A front page *Washington Post* article quoted Judge Green as responding that "We are having problems with Mexican idlers that are not being supervised.... The Mexican consulate should do some constructive work in Chicago and not allow its subjects to be a burden to our city."[11] After Green completed that arraignment, Dominguez went to the front of the court and asked to be heard. Green replied, "I am hearing another case and don't want to hear you."[12] After Dominguez refused to stop objecting to Judge Green's characterizations of Mexicans and his use of the vagrancy law, Green ordered

Dominguez arrested and summarily sentenced him to six months in Cook County Jail.

Not surprisingly, Green's actions caused an international incident with strong protests to the U.S. State Department by the Mexican ambassador to the United States and the Mexican foreign minister. The Associated Press newswire distributed the story and Mexico City newspapers railed against U.S. injustice and mistreatment of Mexicans in the United States. The fact that the *New York Times* and the *Washington Post* also kept close tabs on the proceedings further indicates the seriousness of the incident. In responding to a question about releasing the consul, Green was quoted in a United Press wire story as emphasizing that his actions were legal and could "not be rescinded by any official or government authority" without his consent. Saying that "Dominguez was guilty of flagrant contempt in my courtroom," Green insisted that "no government official, not even President Hoover, can make me back down."[13] Green's brashness and condescension, especially when dealing with Mexicans, might not have been indicative of the general feelings of greater Chicago society, but it *was* representative of ethnic white attitude in South Chicago. Green, only thirty eight years old at the time of this incident, was born and raised near South Chicago by U.S.-born parents and an Irish grandfather. His father worked in construction at a steel mill. His grandfather, who is listed in the 1900 census as living with Green and his parents on the corner of 95th Street and Blue Island Avenue, was born in Ireland and immigrated to the United States in 1851. The likelihood that his parents identified strongly as Irish helps explain his anti-Mexican bias during this period.[14] It took a State Department letter to the Illinois governor to force Green to retract his decision regarding Dominguez. On July 10, after meeting with the chief municipal judge of the City of Chicago at the behest of the governor, Green quashed the contempt charge and expunged Dominguez's record "at the request" of the chief justice. Dominguez went on to an exemplary career serving as head Mexican consul to Houston, Dallas, Sacramento, and Los Angeles.[15]

Although Green's anti-Mexican fervor was not part of an organized citywide deportation effort, Green's summary harassment of Mexicans from behind the judicial bench instilled fear and was, in their eyes, but one example of an institutional bias against the Mexican community by officials and power brokers. As the depression worsened, so did the attacks on Mexicans.[16]

Max Guzman's experiences with car number four and staying far away from the police were common. For Mexicans living throughout the United States, the Great Depression marked an era of forced repatriation in which immigrants and their American-born children were forced onto railroad cars and transported to the Mexican border.[17] However, and despite Judge Green's animus and action against Mexicans in his bailiwick, the depression-era experience of most South Chicago Mexicans was not one of coerced journeys south in railroad cars. Compared to California and even nearby Northwest Indiana, forced repatriation in Chicago was rare, and the institutional bias in Chicago manifested itself in less extreme and systematic ways. The number of Mexicans who left the Chicago area between 1930 and 1934 is estimated to be 7,500, or about 40 percent of the number of Mexicans in Chicago in 1930; the overwhelming number of them left voluntarily or through some level of coercion.[18]

The types of options Mexican immigrants had in the repatriation process, and the choices they were able to make, varied widely and were not always clear. For example, a common technique among relief workers to coerce Mexicans during the depression was to emphasize that they were not eligible for relief if they were not citizens. This meant, according to the relief workers, that their best option was to return to Mexico. Since most Mexican immigrants did not want to naturalize, the "choice" was between remaining without work, money, or a way to eat and returning to Mexico.

This "choice" was not always as cut-and-dried as some relief workers led immigrants to believe. Citizenship requirements for Works Progress Administration (WPA) projects changed over time and were open to local interpretation. A large number of questions and complaints flowed into state and federal WPA offices as late as 1939 because of confusion based on the complexity of WPA programs and their eligibility requirements. According to WPA administrator Francis C. Harrington, in 1939 "any American citizen, *or other person owing allegiance to the United States* [emphasis mine], who is 18 years of age or older, able-bodied, unemployed, and currently certified as in need" was eligible for aid.[19] To demonstrate allegiance, immigrants usually had to file a Declaration of Intent or "first papers."

This declaration was the first official proclamation of one's intent to become a citizen but did not automatically lead to citizenship. Immigrants had to wait at least two years but no more than seven years to file their "second papers," or the Petition of Naturalization. An immigrant during this period could file his or her Declaration of Intent as soon as he or she arrived in the country, but submitting their Petition of Naturalization required five consecutive years of residency in the United States.[20] Harrington went on to state that "State and local practice generally requires legal residence; the WPA itself makes no restrictions," leaving the decision up to local and state officials.[21] To add to the confusion, these requirements changed over time through executive order or federal legislation. By the end of 1939, "owing allegiance" was not sufficient for participation in a WPA program. Applicants had to have citizenship. Before 1936, no restriction existed for those who could prove they had entered the United States legally.[22]

This requirement for proof of legal entry further complicated the process of putting in "first papers." Many Mexican immigrants to the United States had difficulties proving legal entry. For much of the twentieth century to that point, entry restrictions usually exempted Mexicans through either de facto or de jure border enforcement policies. According to Historian Patrick Ettinger:

> Even when political pressures finally forced authorities to try to effectively reduce Mexican legal and undocumented immigration in 1929, their efforts proved dramatically insufficient to the task. Thanks in large part to the benign neglect of Immigration Service authorities in the preceding twenty years, a generation of Mexican workers had become firmly entrenched in patterns of northern migration by the late 1920s. The structural dependence of the American economy on Mexican immigration, born in the economic and political events of this period, created the circular migration patterns that ultimately limited the ability of American authorities in later decades to control cross-border flows. Only the severe labor market crisis of the Great Depression could succeed in bringing under control Mexican cross-labor migration.[23]

Because of the evolving border-crossing requirements and the willingness of border agents to allow Mexican immigrants to enter the country without official sanction, immigrants were not always issued legal entry documents. *Mexicanos* who remained in Chicago focused on survival regardless of their eligibility for WPA relief programs. Working one day every two weeks—if that—and refusing to go on public aid, Lucio Franco remembered buying twenty-four pound sacks of flour for 25 cents then going to the city dump to gather fruit, potatoes, and pork legs in order to survive. Some eligible *Mexicanos* did take advantage of relief programs such as the WPA. According to Alfredo De Avila, Roosevelt's programs put him and other Mexicans to work "on the banks of the rivers, the creeks, the lake, fixing Calumet Park and other places around the lake. We would get $55 cash for five or six days of work. Then we couldn't work again for another month, again for five or six days. With that we made ends meet."[24]

The effects of unemployment hit the Mexican community of South Chicago hard as members struggled to deal with unequal treatment by social workers, law enforcement officials, judges, business leaders, politicians, and others in South Chicago. Some immigrant advocacy groups, including community centers, protective leagues, and other charities, acted as watchdogs against institutionally and organizationally based abuse in South Chicago, but their existence alone did not prevent abuse. Even in the absence of significant forced repatriation drives in Chicago, Immigration Service officials and other authorities worked to create fear of such an eventuality with harassment, raids, and detainments. Guzman's memory of police car number four looking for Mexicans to arrest in South Chicago and the abuses by Judge Green suggest the success of such campaigns. In addition, federal immigration officials based in Chicago frequently raided pool halls with and without local police assistance, harassing and detaining Mexicans until they could prove their immigration status.[25]

Because Chicago institutions and settlement houses had experience helping and working with many different population groups who had settled in the city, including ethnic Europeans, African-American migrants from the American South, and Mexicans, they had an institutional tolerance that was absent in most other American communities with significant Mexican populations. In Gary and East Chicago, Indiana, where a significant level of ethnic tension dated to the Great Steel Strike of 1919, Mexicans were subject to repatriation campaigns organized by the American Legion and funded by local business and political leaders.[26] Active opposition by immigrant advocacy groups, settlement houses, and Mexican groups prevented the same type of forcible repatriation within Chicago. This does not mean that all Mexicans who left Chicago wanted to leave. Citizenship-based restrictions on government relief programs, along with financial incentives by local officials and the Mexican government for Mexicans to return to Mexico, left many members of the Mexican Chicago community with only one choice—return to Mexico.

The ripples of Black Tuesday, October 29, 1929, reached across geographic, class, racial, and ethnic lines. Nationwide, unemployment jumped from six million in 1930 to eleven million in 1932.[27] The Great Depression deeply affected Mexicans throughout the United States. Mexicans regardless of citizenship status, and many Mexican Americans, received the brunt of the popular frustration from a nation that looked to assign blame for lost jobs and the shattered economy. Mexicans were the newest immigrants in South Chicago, the rest of Chicago, and much of the Midwest, and the largest non-white group in Texas, California, and the American Southwest. Governments, private sector organizations, and individuals found convenient targets for their frustrations and anger in the Mexican community. The most visible effects on the Mexican community were the repatriation programs, voluntary and otherwise, that emerged throughout the United States and led to the eventual repatriation of 415,000 Mexicans nationwide. Of the 415,000 persons "repatriated" to Mexico, an estimated 85,000 were American citizens of Mexican descent.[28]

The significant loss of population was the most dramatic effect of the Great Depression on Mexican South Chicago and on all the other Mexican communities in the Chicago area. Seemingly in tandem with the severe hunger and unemployment that affected much of the city, highly organized voluntary and coerced repatriation campaigns led to the dramatic decline in the Mexican population of the city from the peak of 19,632 in 1930 to 14,000 in 1933 to 13,021 by the end of the following year. By the end of the decade, the U.S. Census Bureau counted 16,000 Mexicans in Chicago.[29] In South Chicago, the Mexican population decreased from 4,241 in 1930 to 2,249 in 1934.[30] From 1933 to 1934, the major Mexican Chicago neighborhoods lost 30% of their populations while the total Chicago area lost only 10%. As many Mexicans left the city and many moved to other parts of Chicago, the percentage of Chicago's Mexican population in the original three neighborhoods dropped to 53% in 1934 from 67% a year earlier.[31] Mexicans in Gary and East Chicago, Indiana, were also part of this mass exodus; each town lost about half of its Mexican population during this period.[32]

In the first book-length study of Mexican repatriation during the Great Depression, Abraham Hoffman studied Southern California and estimated that officials forcibly deported over 415,000 Mexican immigrants and Mexican Americans from the United States with another 85,000 Mexican

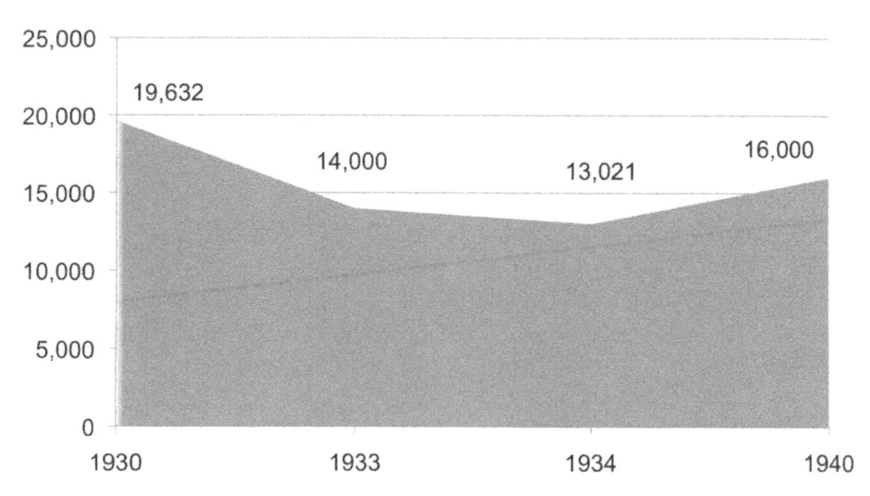

Fig. 7.1. Mexican population of Chicago, 1930–1940.

immigrants voluntarily repatriating.[33] Historian David Gutiérrez estimated that an average of 80,000 Mexicans repatriated every year between 1929 and 1937, with a peak of 138,519 in 1931. Because of the unreliability of immigration and repatriation data, scholars' estimates of the number of repatriates to Mexico during the ten-year depression period range from as low as 350,000 to as high as 600,000.[34] Within these large numbers of "repatriates" were children and young adults who were United States citizens by birth and had never lived in Mexico.

Repatriation drives took place in Chicago, Gary, and East Chicago, Indiana, as well as in California, Texas, and the Southwest.[35] Even after the number of Mexicans returning to Mexico from points throughout the United States peaked in late 1931, officials in Los Angeles, the Chicago area, and Detroit continued to encourage repatriation well into the mid-1930s and the New Deal era.[36] The Gary and East Chicago repatriation drives were so successful that these cities were able to send a majority of local Mexicans to the border much sooner than Chicago or Detroit.

Although East Chicago's large Mexican population and steel industry, and the socio-economic ties between them and South Chicago, made these Mexican communities similar, few similarities existed in how the respective city leaders dealt with Mexicans during the depression. East Chicago had a large, highly organized, and well-funded involuntary repatriation campaign. The American Legion coordinated this campaign in close cooperation with the township's trustee's office and the Emergency Relief Association of the Community Chest while the East Chicago Manufacturers Association contributed financing. Working in concert, these organizations denied aid to Mexicans in East Chicago and blamed them for taking jobs away from members of the dominant community.[37] In a 1932 letter to the Secretary of Labor, an East Chicago American Legion repatriation official ascribed the city's economic hardship on 700 Mexican families living in the community. "The entire employment problem of the county would be solved if they were gone," wrote the repatriation official, Paul Kelly. Not only were Mexicans the cause of economic hardship, argued Kelly, but they also carried feared communicable diseases such as syphilis and tuberculosis. He summed up all of the community's problems by stating that "Herein is our problem, to rid this community of Mexicans." Not only was Kelly quick to blame Mexicans for all the community's past, present, and future ills, but also for good measure he added that Mexicans could not handle the harsh winters.[38]

In Gary, local anti-Mexican feelings culminated during the Great Depression into a government-sponsored "campaign of threats and harassment against Mexican workers."[39] Ironically, the chief spokesman for repatriation in Gary was president of the main Mexican employer, U.S. Steel. Horace S. Norton, also president of the local chamber of commerce and Gary Commercial Club, spearheaded the push to rid Gary of his former employees. Although a voluntary repatriation drive during 1930 and 1931 in Gary provided financial assistance to those wanting to return to Mexico, Gary businessmen, city trustees, and nativists had organized a city-wide, forced-repatriation campaign by 1932.[40]

Mexicans in South Chicago and the rest of the city, with help from allies, developed effective, organized opposition to repatriation drives and the groups who advocated the removal of Mexicans from South Chicago. The Immigrant Protective League (IPL) headquartered at Hull House, *mutualistas, pro-patria* groups, settlement houses (many of which had Mexican employees), and an active Spanish-language press led the way in the protection of local Mexicans during the depression. Although these organizations did not prevent abuse and coerced repatriation, they did prevent large-scale organized repatriation campaigns such as those that occurred in Detroit, East Chicago, and Gary.[41]

Despite the work of progressive organizations in Chicago and the cooperation of the Mexican consul in encouraging out-of-work Mexicans to leave for Mexico, round-ups and deportation drives did occur. An October 1931 *Chicago Tribune* headline read: "Deport 200 in Chicago War on Smuggling Ring."[42] The story mentions the apprehension and questioning of over 125 Mexicans at the South Chicago police station, matter-of-factly stating that three were detained while "others who promised to show their passports were given until today to find them." Not seeing any problem with a mass round-up that netted only three people with immigration violations, the unnamed *Chicago Tribune* writer attempted to justify the round-up of Mexicans by saying, "Other racial groups, it is understood, are to be called in within a few days."[43] Curiously, this operation was conducted by a team of thirty immigration agents from outside of Chicago and led by Murray W. Garrison, a special assistant to the Secretary of Labor. This operation was conducted without the knowledge or prior cooperation of the head immigration agent in Chicago or the Chicago police. This lack of coordination indicated internal disagreement among different levels of government on how to handle "racial" immigrants in Chicago. Garrison's team of agents had been in Chicago for over a month, according to the same *Tribune* article, "to probe the operations of a so called 'billion dollar smuggling ring' which brought aliens into the country without formality and at highly profitable rates."[44] The stated purpose of the team was to go after high-level human smuggling operatives. The reality was that the highest profile round-ups were coordinated raids on South Chicago Mexican pool halls. The federal team used federal support to attempt to do something local law enforcement officials were unwilling or uninterested in doing—namely, engage in high-profile mass round-ups.

In April of 1931, just four months before the Garrison round-ups in Chicago, the *Los Angeles Times* reported that former railroad trainman and current Secretary of Labor William N. Doak "ordered a 'non-spectacular but thorough' drive to deport approximately 400,000 aliens said to be illegal residents of this country." Going after everyone who had "no right to be here" in order to give their jobs to "Americans," Doak equated his deportation drive to a war in which the enemy was illegal immigrants, the majority of whom were Mexican. Secretary Doak argued: "Some may say to deport these people is inhuman, but my answer is that the government should protect its own citizens against illegal invaders. This I propose to do with every weapon in my power." By using military terms such as "weapons" and "invaders" to describe his deportation campaign, Doak attempted to provoke the public to support his campaign out of a sense of urgency and fear. His language successfully agitated the general public to support his campaigns and inflamed tensions among immigrant, African-American, and Anglo-American communities. In emphasizing his "fairness" by going after everyone regardless of ethnicity, religion, or political affiliation, Doak lent credence to the popular idea among unemployed non-Mexicans that aliens had taken their job. Doak boasted that the deportation of "more than half" of 1,100 aliens from New York City in early 1931 "created jobs for unemployed Americans."[45] He promoted the time-tested and still-popular concept that "*they* take our jobs" in times of economic crisis and the deportation of these non-white immigrants will help unemployed "Americans" get jobs.[46]

As mentioned earlier, accurate numbers are hard to come by when dealing with Mexican immigrants who did not trust government officials and entered without record. The unknown number of undocumented Mexican immigrants, along with the migrant living patterns of many Mexican workers in the United States, make it close to impossible to track the number of depression-era departures to the border. To add to the uncertainty of the numbers, some who repatriated—whether they did so unwillingly or took advantage of free or low-cost transportation to Mexico—crossed back into the United States after only a short time in Mexico because of poor living conditions in the *patria*. Scholars have proven that official U.S. and Mexican border crossing records are unreliable and inconsistent because of lax recordkeeping and the fact that many immigrants bypassed official border crossings.[47]

Scholars of Mexicans in the United States have traditionally cited the Los Angeles' Mexican community as the most affected by the repatriation campaigns because of the high level of tension between the Mexican community and Anglo Americans and the negative consequences on the community brought on by the extremely organized repatriation campaigns. The relative newness of South Chicago's Mexican community when compared to Mexican communities in Los Angeles, other parts of California, Texas, and the Southwest; Chicago's distance from the Mexican border; the fact that Mexicans there were not in territory that had once been part of Mexico; and the politically progressive tolerance of immigrants by most city-wide and neighborhood leaders—all these contributed to depression-era changes that might have been different from other communities, but no less dramatic. Hoffman argues that although the largest numbers of Mexicans were repatriated from Los Angeles, Los Angeles' programs differed from those elsewhere only in their degree of organization and duration. While only 3.6% of Mexicans in the United States lived in the states of Michigan, Illinois, and Indiana, more than 10% of those repatriated came from these three states.[48] Several factors make this comparison problematic. One factor that complicates a comparison between Los Angeles and Chicago is Los Angeles' proximity to the border, which made it possible for many voluntary repatriates to enter Mexico without the aid of the repatriation programs, while the distance of the Midwestern communities to the border made the journey to Mexico prohibitively expensive without aid from these programs. Additionally, the large numerical difference in the Mexican population of both regions makes the larger Midwestern repatriation percentages deceptive. These variables also add to the difficulty of comparing the levels of voluntary versus coerced repatriation in California and the Midwest.

The fact that Mexicans left the United States for Mexico in large numbers during the depression is indisputable. The fact that some Mexicans left completely on their own initiative and others were forced onto railroad cars bound for Mexico is also indisputable. Questions arise when trying to delve into the motivations of those who left willingly or might have repatriated unwillingly. How many Mexicans in the Chicago area were coerced by threat of physical violence or denial of relief and relief services? How many of these Mexicans left because they were tired of the discrimination, hardship of unemployment, and harassment and therefore opted for a free trip to the border? How involved were the Mexican consul's office and the relief officials in these repatriation drives? What promises or widely believed promises were understood to have been given by the Mexican government to those who found their way to the U.S.-Mexican border? Despite the fact that involuntary repatriation campaigns in South Chicago were smaller and less dramatic when compared to programs in cities like Detroit, Los Angeles, and Houston, the value of the experiences of repatriates and the impact of their absence on South Chicago are no less important.

Because the vast majority of scholarship on Mexican repatriation during this period focuses on California, and because a significant portion of scholarship examines the rest of the U.S.-Mexico border region, statistical and anecdotal evidence relating to these programs along the border are plentiful. Despite the abundance of Mexican communities that predated the U.S. annexation of the area, Mexican immigrants and Mexican Americans found themselves with little or no political or social power to curb the involuntary repatriation programs. Established, middle-class Mexican-American communities in cities such as Los Angeles and Houston often favored these repatriation programs as they helped to distance the Mexican-American middle-classes from the unskilled and working-class Mexican immigrant communities. In Houston, for example, the *Comité Pro-Repatriación*, a group consisting of middle-class Mexican-American leaders of social and recreational clubs as well as pro-Mexico organizations, raised money through social events and donations to assist the Mexican consul in repatriating working-class Mexicans.[49]

The Mexican government, through its consulates, publicly endorsed repatriation programs while at the same time privately voicing concerns about how to handle such large influxes of people entering Mexico. Mexican government officials, still trying to recover economically from revolution and turmoil, also quietly worried about how the national economy would react to the sudden loss of national income as those without work and repatriating would no longer be infusing dollars into the Mexican economy through remittances.[50] Many Mexicans throughout the country took advantage of subsidized transportation to leave for Mexico, believing that they had been unfairly blamed for the nation's economic woes. Regardless of the level of coercion, many who repatriated felt that they had no choice but to move to Mexico.[51]

Although a sojourner attitude continued to persist within the Mexican community, repatriation advocates forced some members of the community to unwillingly return to Mexico. Many who retreated to Mexico had originally dreamt of returning only after earning enough money in the United States to buy land and provide a comfortable living for their family that would improve their family's economic and social standing in their rural communities. Since these immigrants' aspirations became distant fantasies during the time of economic crisis, those who left harbored hopes of a better life among the stronger familial and social networks that awaited them upon their arrival in their Mexican hometowns. These optimistic ideals of a better everyday life in Mexico remained alive among many preparing to repatriate to Mexico despite their keen awareness of Mexico's fragile national economy and the Mexican government's struggle to accommodate the flood of *paisanos* arriving at the border.

Because of the relative absence of involuntary repatriation programs in Chicago, many Mexicans in the Windy City viewed repatriation as the best, or only, option for individual and family survival. As a social worker, community leader, and advocate, Mercedes Rios participated in what she considered voluntary repatriation of Mexican families. Rios believed that repatriation was a way out of misery for those unemployed Mexicans who might find familial and community support back home. Many immigrant advocates in Chicago shared her attitude.

The Mexican government at times aided repatriates in getting to the border and then to settlements in the Mexican interior. In her 1980 interview with Jesse Escalante, Rios recalled hearing about a 1933 letter from the President of Mexico that promised any poor family that wanted to return to Mexico free land and transportation from the border to their new land.[52] A letter from the Mexican consul to Chicago, Rafael Aveleyra, to the editor of Chicago's *El Nacional* newspaper published on May 14, 1932, and dated four days earlier, verifies Rios's claim of the Mexican government's involvement in voluntary repatriation of Mexicans in Chicago. Aveleyra informed Mexicans in Chicago that "through the efforts of the Secretary of Foreign Relations" it was possible for Mexican families to return to Mexico at a subsidized cost of $15 via the Missouri Pacific Railroad.[53] In a follow-up story by the editors of *El Nacional* published two weeks later, the program was declared a success with "as many as fifty Mexicans, many of them with their families," taking advantage of this specific voluntary repatriation program. In addition to the $15 fare reported in the letter by Consul Aveleyra, the Mexican National Railroad and the Mexican Government were cooperating in fixing "transportation rates at one cent per mile for those who will travel from Laredo to points in Mexico." The newspaper article also reported that the news of subsidized transportation on both sides of the border was met "with great approval by the Mexican Colony of Chicago." It was the reporter's opinion that many "countrymen" would undoubtedly "take advantage of this opportunity."[54]

With the promise of a better life in Mexico, Rios asked "quite a number of them when they came in for their checks ... if they wanted to go to Mexico."[55] Although it was common practice in other cities such as Los Angeles, Detroit, East Chicago, and Gary, Indiana, to deny relief in order to force Mexicans onto border-bound trains, evidence suggests, as does Rios, that South Chicago Mexicans did not experience this form of coercion. Rios recalled that many said that they wanted to return to Mexico, "but they had no funds." She claims to have helped about twenty-five large families who agreed to return to Mexico and went as far as walking them to the train station. In addition, she made "sure that they got money and enough food to go to the border."[56] Rios's memories of repatriation highlight the complicated nature of voluntary and coercive repatriation and the motivations behind those who advocated it. In industrial cities like Chicago that were far from the border, voluntary repatriation through one of the established programs provided a "free ride" home and promises of a helping hand from the Mexican government once the immigrants arrived at the border. With limited options in Chicago, many left on their own accord with the blessings of social workers and immigrant advocates. On the other hand, immigrants might have perceived the advocates' active endorsement and "encouragement" to go to Mexico to be coercive in nature.

Alfredo De Avila also remembered repatriation as an alternative to welfare in South Chicago. He recalled that "Many of the families, many of them my friends, decided it was best to go to Mexico."[57] Serafín García agreed that "in South Chicago they mostly were volunteers." He explained that "They didn't force nobody. I did hear that in East Chicago, Indiana, many families were forced to go back. But not here in South Chicago." His family stayed because his mother felt there was nothing to return to in Mexico, as his mother's family and her husband had died.[58]

As late as 1936, repatriation remained an important issue for Mexicans in Chicago. An article in *La Defensa,* reported on discussions by the Mexican government to solve "the problem of repatriation." The newspaper reporter explicitly stated that he expected many Mexican immigrants living in Chicago to return to Mexico should this later Mexican government repatriation program materialize. In addition to the two million pesos appropriated to help repatriates, the article mentioned that since the onset of the depression, Cook County social agencies had paid the expenses of "many Mexican families to the Mexican border."[59] Antonio L. Schmidt, Chicago's Mexican consul in 1937, stated that eighty-three Mexican families repatriated from Chicago in 1936. He emphasized that only those "carrying with them certain instruments, such as radios, washing machines, typewriters, or any merchandise that need to be checked by our office to ... verify the documents which may show its legal acquisition," were required to register before traveling to Mexico. The families that registered as repatriates, he recalled, ranged in size from two to five with only a few exceptions. With only those needing to declare items registering, it is likely that at least two to three times the number of families mentioned by Schmidt actually repatriated that year.[60]

One Mexican organization active in South Chicago that worked to protect Mexicans already in the repatriation process was the *Frente Popular*. The Chicago committee of the *Frente Popular,* loosely affiliated with the Mexican organization by the same name, was established in January 1936 just after the first U.S.-based committee was established in Cleveland, Ohio.[61] During meetings of one of the local branches of the *Frente Popular* held in March 1936, leaders emphasized that "One of the main ideas in [the organization's] platform is to protect the Mexican who returns to his own country in the repatriation groups." The group advocated for guarantees that those returning to Mexico from the Chicago area "be given land, water, and farm implements, so they will be able to re-establish themselves in Mexico."[62] They emphasized the fact that without guarantees by the Mexican government, repatriation was a dangerous option.[63]

According to notes taken by a University of Chicago Settlement House worker during a meeting of the *Frente Popular's* Back-of-the-Yards branch, "the speaker spoke strangely in favor of repatriation guarantees which would safeguard the returning Mexicans so that they would find good conditions in the home country. They want land, water and tractors absolutely guaranteed by the Mexican government so there will be no danger of the home comers finding themselves in a situation more miserable than the one they left in the United States."[64] What must have seemed strange to the social worker was the fact that members of this Mexican community organization spoke in favor of repatriation and looked for guarantees within the repatriation process, instead of fighting against repatriation, at least in principle. It is worth noting that the *Frente Popular,* or Popular Front, was eventually considered a communist organization by settlement house leaders and kicked out of the University of Chicago Settlement House. Linking communism and the *Frente Popular* was not difficult to do, given the organization's labor activism, links to more radical leftists, and the *Frente Popular* of Mexico's loose ties with the Lazaro Cardenas administration and the Mexican Communist Party.[65]

Many South Chicago Mexicans who stayed in the neighborhood ended up accepting charity or public aid to survive. Survival required them to swallow their pride and take part in depression-era relief programs. The crisis served as an ongoing reminder that they were not equals in the eyes of those around them. According to Illinois state officials, 778 Mexicans from 186 families living in South Chicago enrolled in relief during 1933. Just over 30 percent of all Mexicans city wide, or 4,272 Mexicans (939 families) out of a total population of about 14,000, were on the official relief rolls.[66]

Mexicanos, including Mexican South Chicagoans, benefited from the many relief programs such as the Works Progress Administration (WPA), the Civilian Conservation Corps (CCC), food and coal distribution programs, direct financial assistance, or "the soup line by the 92nd Street bridge."[67] In the eyes of Rafael Guardado, Anthony Romo, José Cruz Díaz, Mercedes Rios, and other South Chicagoans, the hardest years of the depression were 1931–1933, when many families felt they had no choice but to go take advantage of government relief programs.[68]

Many Mexicans in South Chicago resisted Americanization, yet the depression forced them to interact with agencies they would have otherwise avoided for fear of being forced into some type of Americanization programs. Anthony Romo could not recall when his family first participated in relief programs, but he remembered that they did participate. Rafael Guardado remembered that, as a single man, he could not participate in most aid programs. Mercedes Rios and José Cruz Díaz remembered the long lines and recalled helping Mexican South Chicagoans complete aid forms. Others remembered the option between participating and/or being given a free train ticket to Laredo, Texas, a city on the Rio Grande and "Gateway to Mexico." For others, relief came through participation in work programs.[69]

The Mexican community of Chicago produced and nurtured leaders from within to provide direction and voice for community members during the crisis. While many from outside of Mexican Chicago considered the Mexican consul to Chicago the voice and leader of the Mexican community, most Mexicans in Chicago instead saw the consul as a representative of the Mexican government, but not a community leader. Because of this official position, local Mexicans—especially Catholics—did not trust the consul to serve their best interests. The consulate was a place to register marriages and births, and a place to seek help and advocacy in dealing with American institutions. Since Mexicans considered the consul merely a government representative, his presence did not inhibit the development of leaders within the South Chicago Mexican community during the Great Depression. Mexican men and women in South Chicago assumed leadership positions to represent and advocate for the community to employers and government officials. Mercedes Rios, who was praised by José Cruz Díaz at the start of this chapter, was one of these depression-era leaders. In addition to her official duties and unofficial advocacy, Rios served as an officer in the South Chicago chapter of the *Cruz Azul*, a depression-era aid organization that originated in Mexico.[70]

Members of the community formed several groups—or local chapters of larger groups—to help other Mexicans in South Chicago to survive and persevere. One of these organizations was the South Chicago chapter of the

Cruz Azul Mexicana, where Mercedes Rios served on the executive board. Rios served at one time as the vice president of the city-wide "Brigade" of the *Cruz Azul* and as secretary of the South Chicago auxiliary.[71] The *Cruz Azul Mexicana* was run by women and had "brigades"—as the local chapters were called—in Mexican communities throughout the United States. Named after an organization operating in Mexico, the *Cruz Azul Mexicana* was actively promoted by Mexican consulates throughout the United States as an organization that could help improve the health and welfare of Mexican immigrants. In Chicago, the support of the Mexican consulate was such that the president of the local brigade was Milla Dominquez, wife of the Mexican vice-consul who was thrown in jail by Judge Green while he was acting consul. Milla Dominguez was also a popular Mexican singer.[72]

The establishment of a South Chicago auxiliary of the Chicago brigade of the *Cruz Azul Mexicana* is itself evidence of a strong Mexican community in South Chicago that was able to fight on its own behalf. Brigade leaders established the South Chicago organization in May 1930, after editors published complaints in local Spanish-language newspapers about the limited relief and aid services the Chicago *Cruz Azul Mexicana* provided to Mexicans in South Chicago.[73]

In an article published in the *Mexico* newspaper, Narciso González, the South Chicago bandleader and steelworker introduced in chapter 6, argued for more *Cruz Azul* attention to South Chicago's needs and more support of the organization by Mexicans throughout the city. As one of its primary services, the organization sent volunteer doctors and nurses into the homes of Mexicans in need of medical attention. The organization's volunteers complained of the time, the expense, and the difficulty of getting to South Chicago from the Near West Side where the organization was based. González called for financial cooperation from Mexican merchants and "the colony in general."[74] In seeking financial support of the organization, González reminded Mexicans that it was the "duty of every good Mexican to be vigilant" over other Mexicans who were "dispossessed and wandering through the streets, nearly begging some bread, not for them, but for their children who cry because of a lack of food, without shelter, or in bed sick, fighting between life and death without any protection or resource [other] than the aid of our charitable institution."[75] Although the dramatic style of the editorial was not unusual in Spanish-language newspapers, this piece highlights the dire circumstances of Mexicans in Chicago and the willingness of community leaders to put out a call to action that emphasized cultural unity and the need for community-based help during the crisis.

In May of 1932, *El Nacional* published a sensational example of the suffering by Mexican immigrants in the Chicago area. The story announced the death of Concepcion Guiterrez de Arriaga who lived in the "slum section of this city." Investigating a six-week-old infant's constant crying, Manuel Castillo entered the building and "stood horrified at the sight of the baby making futile efforts to reach its mother's bosom. Her body was motionless." The article went on to describe the scene in detail: "Five small unclothed and hungry children were standing around the bed where their lifeless mother was stretched out. The windows of the house were boarded up, leaving the place in complete darkness." The article explained that the children's father had abandoned them months earlier, and that

> on various occasions they spend days without a piece of bread and that the sacrificing mother did not desire to give her children to a charitable institution for fear that they would not have the care which she could give them. She preferred to have them at her side in misery and poor surroundings, doing what she could to give the food, which their heartless father had denied them.[76]

The mother, according to the paper, died of starvation even as "on the table of that room, practically unfurnished, was a piece of hard bread, probably the last contribution the mother had obtained for her children and which had as yet not been given to them for fear of increasing their hunger."[77] To further illustrate the tragedy, the paper listed the children by name, as well as one other who was in the hospital with a broken back from a recent fall out of the apartment's second floor window. The children found at home were eleven, seven, five, one, ten months, and six weeks.[78] Although starvation and this level of tragedy was not common in Chicago, the lack of nutrition and fear of aid agencies was real and widespread.

In his call for financial support of the *Cruz Azul* to provide better services for the Mexican community of South Chicago, González argued that the only way to raise enough funds to support the organization was through festivals. Writing that the idea was "magnificent," González argued that recreational festivals, as well as those held "in commemoration of the struggle and deeds of our Mexican heroes," would solve the organization's financial woes only if members of the "Mexican Colony in Chicago and vicinity" followed through by freely contributing their "share every time they [were] solicited." He suggested activities that not only helped the community by raising funds for an important organization, but also strengthened pride in that community through cultural celebration, recreation, and the sense of coming together to help other members of the community.[79]

In response to the above article, South Chicago physician Dr. Oscar G. Carrera, a former *Cruz Azul* executive council member, published an open letter to the executive board of the Chicago brigade of the organization. In it, he called for the creation of a semi-autonomous South Chicago sub-brigade or delegation. Carrera argued that because Narciso González was a working-class member of the South Chicago community and was "in close contact with both the existing misery and prosperity in the Mexican colony of South Chicago," he was "authorized to speak in the matter the way he did." Additionally, Carrera argued that the article was highly significant because it had been written by "a humble person, a laborer," and was an example of the high level of interest for the *Cruz Azul* among the Mexican working class in South Chicago.[80]

By May 6, one month after the publication of González's article, the *Cruz Azul's* executive board formed the South Chicago Auxiliary. The importance of the auxiliary to Mexicans in South Chicago was clear in the festival-like "literary-musical function" and meeting held to choose the officers for the new organization. Over 400 Mexicans attended the event that was hosted by Consul Aveleyra. In addition to providing medical attention, the South Chicago *Cruz Azul* coordinated community-wide efforts to collect cash and supplies to help needy Mexicans in South Chicago.[81] Of note is the lack of official cooperation from Our Lady of Guadalupe Catholic Church because of the *Cruz Azul's* active sponsorship by the Mexican government.

Another national organization with a city-wide presence that had a branch in South Chicago was the *Frente Popular*. Officers and members of the *Frente Popular* of Chicago and of its community-based branches also sought ways to help local Mexicans in need during the Great Depression. Aside from their advocacy to improve the plight of those returning to Mexico, the *Frente Popular* worked to create programs to help Mexican immigrants and provide an avenue for the maintenance of Mexican culture in Chicago. Through the work of each community-based branch, including the one in South Chicago, these programs helped in the formation of strong depression-era communities in Chicago.

On February 2, 1936, an organizational meeting for the city-wide committee of the *Frente Popular* was held at Hull House and brought together members representing the South Chicago, Back-of-the-Yards, and Near West Side branches of the organization as well the general Mexican public. The Sunday afternoon meeting, meant to "talk over general plans for the Committee and to promulgate the ideas contained in the platform," was attended by an overflowing crowd, with at least fifty people left to stand the entire meeting. The attendees were "composed of all classes, from the educated and professional group to laborers, the latter outnumbering the former in a ratio of about twenty to one" and included a "liberal sprinkling of women both young and old."[82]

One of the event's speakers, an articulate young Mexican man, "pointed out that a great many Mexicans who had come to the United States to ameliorate their conditions now find themselves in a much worst state both socially and economically than they were in, in the home country." The Mexican immigrant to Chicago, he argued, had left the "cultural advantage of their homeland" and had been, in most cases, "unable to avail themselves of American cultural advantages." To meet the needs of members of Mexicans in Chicago, the leaders of the local *Frente Popular* pushed for the development of three projects. First, they argued that libraries of Mexican books should be made available to all Mexicans in Chicago. To that end, books sent to the organization by the Mexican Department of Education were distributed to the community-based branches for public access. Second, the group called for the establishment of Mexican schools in Chicago to be financed by the Mexican government. The purpose of these schools, they argued, would be to "transmit Mexican culture to the younger generations." Lastly, the group called for the establishment of scholarships for "talented, young" Mexican Chicagoans to go to Mexico and train "in the arts and professions." These scholarships would help young members of the Mexican communities in Chicago "become leaders of their own people both here and in Mexico."[83] The goal of all three of these initiatives was to create strong Mexican communities and develop educated leaders from within these communities who would advocate for the continued promotion of Mexican culture and Mexican cultural identity within their communities and to the dominant society in Chicago.

Mexicans who remained in South Chicago during the Great Depression helped advocate and provide for members of their community. The relative newness of the Mexican community to the area, the prejudice displayed against them, their sojourner attitude, and their resistance to naturalization and Americanization were all factors in defining why and how South Chicago's Mexican community developed and reacted as it did during the Great Depression. Although immigrants turned to the Mexican Consul for advocacy and assistance, and applied for public relief, leaders emerged. The actions by individuals and the organizations they led came from a sense of obligation that members felt for their community. Although the Mexican community of South Chicago emerged from the depression much smaller, the sense of community among Mexicans in South Chicago was stronger as they organized in order to survive harassment and the depression itself.

8

Teamwork

Men in South Chicago, like Serafín García, created and joined sports teams during the Great Depression to be able to get out of the house, to socialize, and to stay active. García remembered having a full life even when unemployed, playing baseball in the summer and basketball in winter. Manuel Bravo, an out-of-work steelworker, concisely summed up his experience during the depression: "There was nothing, no work, no nothing. The only recreation was playing baseball and more baseball, basketball and more basketball. So we turned out a lot of great baseball players and basketball players."[1] Organized sports were clearly a central part of everyday lives for unemployed Mexican men during the Great Depression and were not "just a game" for those playing, those watching, and the Mexican community as a whole.[2]

Organized sports were much more than a way to cope while waiting for a few hours of work at the mill, your turn to pick up a few days of work in a WPA program, or your turn to collect relief. Youth and adult baseball provided a much-needed distraction for Mexicans in South Chicago who faced the social, political, environmental, and economic discrimination outlined in previous chapters of this book. Sports provided youth like Gilbert Martínez, Serafín García, Pete Martínez, and Angel Soto the opportunity to represent their community within the neighborhood or around the area. For Mexican business owners and other community leaders, baseball became an opportunity to instill pride, teach, celebrate, and organize the community.

The men and women who played or organized teams were important actors in shaping Mexican South Chicago's physical environment and in improving individual and community welfare during the Great Depression. Players may have been struggling to survive, but when they put on a uniform and competed against teams from outside of South Chicago, they not only represented their community, but also contributed to it through "leisure" activities that were much more than a distraction from the horrible realities of being part of a marginalized population in a polluted urban environment during the time of crisis. This chapter examines the development of these teams and their leaders from organizers of recreational activities to founders of organizations that helped coalesce a physical and imaginary Mexican South Chicago identity and provided a means of interaction with other communities in and around Chicago. In examining the influence of the teams, I examine methods used by community members to shape their environment in order to survive stress and the swirling pressures linked to the ever-present and increasing harassment, discrimination, and repatriation drives that "encouraged" some to leave for Mexico.

South Chicago Mexicans used recreation, primarily organized sports, not only to persist and persevere, but also as a vehicle to create organizations that promoted positive cultural and physical environments, which in turn improved everyday life for members of the community. Important to the community's survival during the Great Depression, these organized leisure activities and the organizations originally formed to support the teams became catalysts in the formation of community among *Mexicanos* in South Chicago. They were cogs in the community's ability to effect change in their environment. People created a sense of community and pride for their neighborhood by participating in organized recreational activities. Leaders who emerged from these organizations not only improved access to parks and recreational activities, but also created opportunities for Mexicans in environments off of the athletic field or court. Community members used these opportunities—some small, others significant—to create pride in their community and persevere in the shadow of the Great Depression.

Individuals in South Chicago, including grocery-store owner and Yaquis sports club founder Eduardo Peralta, newspaperman and newsstand owner Eustebio Torres, and steelworker Serafín García developed into community leaders and activists through their involvement in organizations that were originally formed to raise money and build enthusiasm for local amateur baseball and basketball teams. Contemporary scholars who studied Mexicans in Chicago during the interwar period assumed that such internal leadership did not exist—or need to exist—because of a strong consular and settlement house presence. Although settlement house workers played an important role in the Near West Side and Back-of-the-Yards Mexican communities, the lack of a large, secular settlement house in South Chicago placed a greater burden on community members to lead from within. As discussed in chapter 7, although the Mexican consul to Chicago was active within the community, the relationship between community members and the Mexican government's representative in the city was not always as close as outsiders believed.

Leisure-time activities were important parts of surviving the Great Depression. Contemporary scholars such as sociologist Jesse Steiner believed that the federal government needed to take a hands-on approach to develop wholesome leisure and recreation choices for unemployed and underemployed members of the middle and working classes. Steiner, a member of President Hoover's President's Research Committee on Social Trends, argued that this "problem of leisure" required the government to intervene and steer people away from "undesirable" activities and towards "wholesome" ones that would be popular with the masses. Scholar Susan Curell emphasizes that the problem of leisure was the "central cultural issue of Depression America." People needed to spend their time in a "healthy" way. Society's ills—argued many reformers, doctors, social workers and educators—were the principle byproducts of "improper leisure."[3]

Historians have defined leisure activities, which include sports and recreation, as those providing personal satisfaction and pleasure, rather than those done for utility.[4] Such a simple dichotomy fails to account for the range of responsibilities among Mexicans and thus their varied choices about time away from the workplace, particularly during periods of unemployment. The separation between work and leisure for women, both mothers and older female children who often assisted in the care of siblings, was not as transparent as it was for men. Although many women who worked outside the home lost their jobs during the crisis, their household duties remained the same or increased as men lost work. These gendered responsibilities, along with cultural mores that circumscribed adolescent girls' and unmarried women's movements to limit unsupervised contact with unrelated boys and men, restricted women's opportunities to participate in organized leisure activities.

The examination of recreation, like other forms of leisure activities, is a useful window into the everyday lives of working-class immigrants and can reveal how participants—both active and bystanders—contributed to a sense of community. In the case of Mexican South Chicago, the analysis of these activities provides insight into how immigrant and U.S.-born adults and children reacted to pressures brought by groups both hostile and friendly to force them to assimilate, or "Americanize." Employers and city leaders attempted to control and contain *Mexicanos* to specific fields of employment and geographic areas of the city while attempting to Americanize them. In addition to keeping *Mexicanos* within certain community and workplace niches, members of the Anglo community sought to control the recreational time and space by trying to keep them within specific recreational boundaries.[5]

Park and community-center staff and volunteers who came from outside of the neighborhood primarily organized youth sports. Leaders from within the Chicago-area Mexican community—such as Eustebio Torres and Eduardo Peralta—organized the men's athletic teams and leagues, rarely working in concert with the outside agencies that promoted youth activities. Under local leadership, men's organizations that began as baseball or basketball teams quickly grew into multi-faceted associations with influence and importance far beyond the baseball diamonds and basketball courts. These organizations had a significant direct and indirect influence on the welfare of the community. In investigating interwar leisure-time activities such as organized sports and social gatherings, several things become clear. First, Chicago-area Mexicans, including those in Northwest Indiana, were not isolated from each other. Second, leaders emerged from within Mexican South Chicago to create organizations that improved quality of life and advocated for their constituents' needs and concerns. Third, despite the rigors and drudgery of workplace and home labor, leisure-time activities played a central role in *Mexicano* life.

Before delving any further into the fascinating and "wholesome" world of organized sports and the organizations they spawned, a brief examination of the role of pool halls in the Mexican community will help us put these activities in proper perspective. Community insiders and outsiders alike considered men's time in the pool hall one of the most pervasive "ills" in Mexican South Chicago. To many, these establishments were much more than places to play pool. In examining reaction to Mexican pool halls patronized by agricultural workers in Corona, California, historian José Alamillo found that "critics objected not so much to pool playing itself, but to its association to vice and criminality."[6] These establishments served the same space as saloons had in industrial working-class America. With very few exceptions, these were masculine spaces where Mexican men could congregate and create community through socializing, playing pool, drinking, and gambling. Men could also engage the services of a prostitute, get into a brawl, and rent a room for the night. Several pool halls in South Chicago had attached restaurants. One even had a barber shop. According to Robert Jones and Louis Wilson, there were over fifty Mexican pool halls in Chicago's Mexican neighborhoods in 1931. They considered the pool hall as being "quite as much patronized as a social center, news dispensing agency, and mutual aid society as for its more obvious purposes."[7] They added that the pool hall proprietor was one of the most trusted neighborhood bankers for working-class Mexican men. In the Millgate section of South Chicago, pool or billiard halls constituted thirteen out of a total of thirty-five businesses. Regardless of what pool hall activity the patron engaged in, it usually included drinking large amounts of alcohol.

The pool hall is where men went to cope with the effects of the depression, discrimination, poor working conditions, family troubles, and other travails. Often, problems came hand-in-hand with the excessive drinking. In addition to addiction, the drinking had immediate effects on family finances, on domestic relationships, and on workplace performance. Alamillo sees the pool hall as a place where "the mix of extramarital relationships, excessive drinking and gambling, and availability of deadly weapons could easily lead to male-on-male violence and violence against women."[8] In addition to poolroom violence by patrons, it was not unusual for police to harass patrons.[9]

When Mexicans were allowed access to parks, those spaces became alternatives to the pool hall for some. Chicago's extensive public park system played a fundamental role in providing Mexicans with amenities, such as shower facilities, recreation, and recreational space for adults and families, including organized, supervised activities for the community's youth. A considerable amount of time passed before other ethnic groups—and in some cases park staff—allowed *Mexicanos* free and unrestricted access to area parks. The all-too-common ethnic "turf wars" throughout South Chicago and the racial and ethnic discrimination that existed within the park system limited Mexican access to parks and recreational facilities through the early 1930s.[10]

The lack of park access was a significant impediment for young Mexicans. In a 1928 study on the Chicago park system, Marian Osborn asserted that play and recreation were critical for adults as well as children. She argued that parks were not only a character-building necessity for children, but also an important outlet for adults. Children's play was more than just "natural and necessary," she maintained; skillfully supervised and directed play was "one of the most powerful agencies in character building." Access to parks and recreational activities was also important for adults who needed diversion from the "drudgery and drabness" of the industrial workplace.[11]

Although public parks are best known for open fields and play equipment, many Mexicans found Bessemer Park through necessity. While growing up in a family of six children in South Chicago during the 1920s, Anthony Romo, along with his family, utilized the park's shower facilities as more than a practical amenity. The time and trouble involved in attempting to warm up bath water for six children on a kitchen stove made the park shower facilities indispensable. By the late 1920s, Mexicans, primarily adults, were using Bessemer Park shower facilities at a rate of seventy-five per day in the summer and fifteen per day in the winter. By 1927—despite

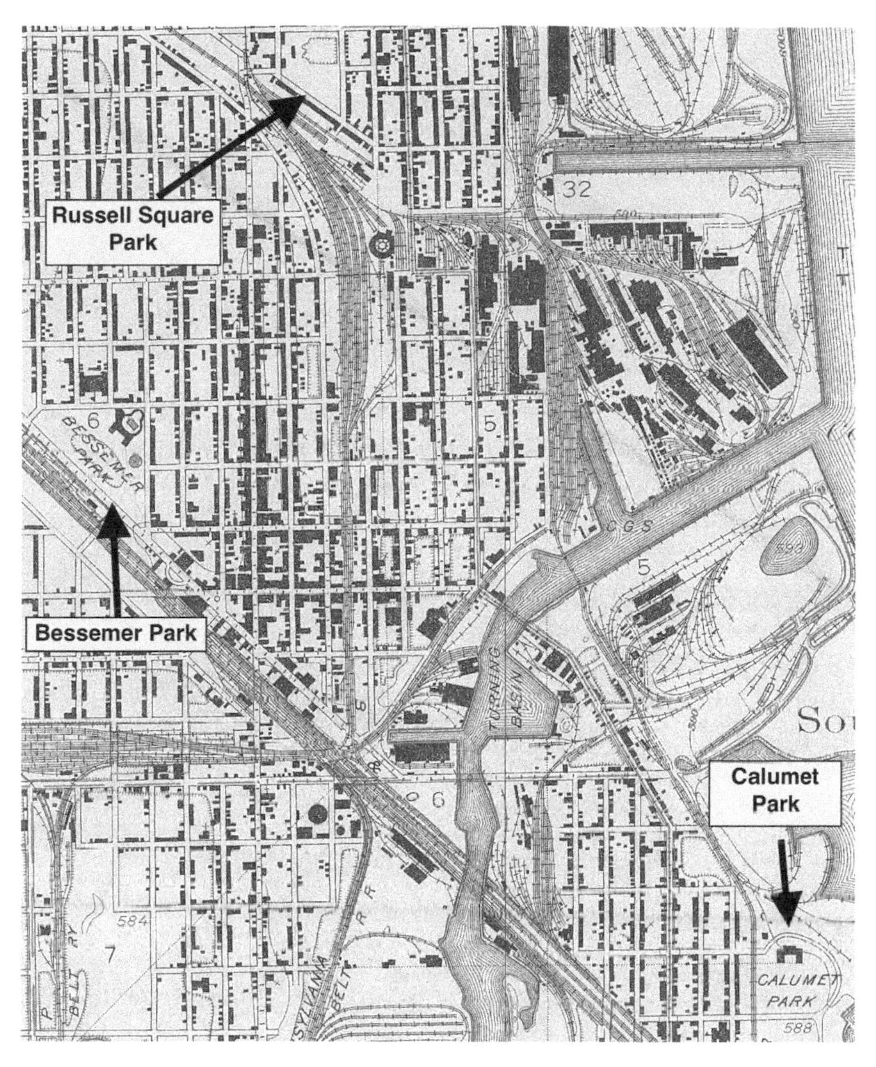

Fig. 8.1. Neighborhood parks. Map courtesy of the University of Chicago Library's Map Collection, home of the original. *7.5 minute series (topographic): Chicago and vicinity.* Washington, DC: Geological Survey; Urbana, IL: Geological Survey Division, 1928–1929.

inter-ethnic hostility—Chicago-area Mexicans were using Bessemer Park for organized Mexican Independence Day celebrations and for the staging of Spanish-language plays in the field house auditorium. Yet, records show that only two Mexican children participated in organized park activities in 1927. The park director reasoned that the lack of youth in park activities was because the facility was too far from their homes. However, the primary reason that young Mexicans avoided Bessemer Park in the 1920s was intimidation by youth of other ethnic groups. Gilbert Martínez recalled that he and his friends would frequently "get beat up by the guys that would hang around there, the *Polackos* I guess, I don't know who the hell they were." These youth, Martínez explained, came from "around from where the police station" was and controlled the park. The police station was at 2938 East 89th Street, only two short blocks from the park. As in Bessemer Park, Spanish-speakers were not welcome at Calumet Park throughout the 1920s.[12]

During the Great Depression, parks became a "healthy" escape from the economic crisis for unemployed adults and their families, thus increasing the Mexican community's determination to use the parks despite hostile attitudes by park officials and others who frequented the parks. For many, access to the parks, whether for team sports, family outings, or simply an evening stroll, was important for their physical and psychological survival. During the height of the Great Depression, Mexicans used Bessemer Park more than any of the other South Chicago parks. Bessemer's facilities included a field house, gymnasium, auditorium, outdoor swimming pool, wading pool, playground, athletic field, and tennis courts, as well as a branch library and a toy lending service. Other area parks frequented by Mexicans included Russell Square Park, Calumet Park, and several facilities in Northwest Indiana.[13]

South Chicago's community centers did not have the human, social, and financial resources of the Near West Side's Hull House or the Back-of-the-Yards' University of Chicago Settlement House. In a 1928 study, Anita Edgar Jones argued that, unlike the two other major Mexican communities in Chicago, that of South Chicago was left to its "own organization" when it came to recreation. The religious affiliations of South Chicago's community centers were also important factors in determining the resources available to them, relative to other Mexican neighborhoods. Though not entirely left to their own devices as suggested by Jones, members of South Chicago's Mexican community played a more active role in organizing and promoting recreational activities than did members of the other Mexican neighborhoods due to the inequality of resources.[14]

South Chicago's community centers were sponsored by Protestant churches and served the Mexican population to varying degrees. The Bird

Memorial Congregational Church's South Chicago Community Center (SCCC) was located at 9135 South Brandon Avenue, only a stone's throw from Our Lady of Guadalupe.[15] The SCCC, commonly referred to as "Bird Memorial," was established in 1923. It was physically attached to the Congregational Church and financially supported by the Chicago Congregational Union. This fact eventually led the Catholic Church to ban the faithful from utilizing the center's gymnasium and activity rooms. Another important institution that was also off-limits to Catholics was the Baptist-funded South Chicago Neighborhood House, at 84th Street and Mackinaw Avenue. Priests and nuns told Mexican youths that using the facilities or participating in organized activities at the above institutions and at the Y.M.C.A. was sinful because of the institutions' Protestant affiliation.[16]

South Chicagoan José Cruz Díaz recalled being confronted by Father James F. Tort, the Catalonian-born parish priest at Our Lady of Guadalupe Church, for using the recreational facilities at Bird Memorial: "Father Tort said that it would be a mortal sin if we went to the congregational church." Tort defined going to the South Chicago Community Center not only as a sin, but also as a major sin that would put an individual's soul at risk. Regardless, Cruz Díaz continued to go to Bird Memorial to play organized basketball for teams such as the Elks and the Mayas because of the facilities and because he felt he was treated "awful nice." "In fact," he added, "we had the run of the place." With the high level of ethnic tension and discrimination at public parks and other area facilities, the fact that Mexican teens "had the run" of the recreational facilities at Bird Memorial was indeed significant. Cruz Díaz understood that Father Tort and church officials were trying to keep them away from Bird because "hanging around the Protestants was a sin" and the church worried about losing members to Protestantism, but Cruz Díaz didn't see the logic in such a policy because he and his friends had little, if any, contact with the pastor and believed that Bird Memorial's motive was simply to keep them off the street. Despite the fact that many Mexican youth and families adhered to the church's order, many *Mexicanos*—Catholic as well as Protestant—frequented Bird Memorial because of its resources, the staff's outreach into the Mexican community, the comfortable, non-discriminatory atmosphere of the center, and a lack of comparable Catholic-run facilities in the neighborhood.[17]

In contrast to Bird Memorial, Common Ground, under the direction of Reverend Raymond Sanford, established and maintained a cooperative relationship with the Catholic Church despite Common Ground's links to the Congregational Church. Unlike the city's public parks, Bird Memorial, and the South Chicago Neighborhood House, Common Ground was an organization without its own recreational space. Sanford ran the organization out of a small office at 3029 East 91st Street. As general director of cultural activities of the local Congregational Churches, Sanford was given $250,000 by the Chicago Congregational Union "to erect buildings to start and carry on this constructive cultural work" within South Chicago. Instead, Sanford developed a sort of "welfare clearing house" for members of the community and worked among the various racial and ethnic groups in the neighborhood.[18]

Sanford's organized recreational activities for neighborhood youth included field trips to local museums, libraries, stock shows, automobile shows, the Adler Planetarium, the Chicago Historical Society, and other "events of importance [that] occurred during the year, including ball games." Sanford gained access to seven Catholic churches as well as most Protestant facilities to use their recreational halls in order to accomplish Common Ground's primary mission: human-welfare work. Common Ground used whatever was available for recreation. This included "old buildings, woodsheds, garages, etc."; on rainy days "some of the activities [were] carried on under the sidewalks which are very much in the nature of tunnels."[19]

With the Works Progress Administration financing much of the equipment and personnel needed for their recreational activities, Common Ground organized many youth activities, including basketball, baseball, football, soccer, chess, checkers, photography, and crafts. WPA workers and University of Chicago students played considerable roles within pre-existing organizations in planning and supervising youth activities throughout South Chicago.[20]

Leisure-time activities were heavily gendered, with Mexican women and girls facing many more obstacles and hardships then men when it came to actually participating in them. Socials hosted by girls' clubs at Bird Memorial, although supervised by staff members, provided a venue for adolescent boys and girls to meet without parents or other family members being present—something frowned upon within the more traditional Mexican circles. The same cultural expectations within the community that facilitated limitations that husbands and parents put on the type of work that women and girls could perform also placed limits on their social behavior. Even as much of the morality-based restrictions placed on Mexican women and girls came from within the community, South Chicago Protestant missions played a role in shaping female "morality" through their neighborhood-based recreation and Americanization programs.[21]

In Mexican South Chicago, like elsewhere, many parents tried to control their daughters' level of "Americanization." Parents believed that with

Americanization came increased independence that would lead to greater opportunities for women's social and sexual autonomy. Participation in social and leisure-time activities did undermine familial control by creating a youth culture where daughters explored romantic relationships away from the "watchful eye" of parents and relatives. Parents also attempted to regulate co-educational interaction in schools and during almost any type of leisure activity. The earlier example of Socorro Zaragoza, an elementary school student in nearby Gary, Indiana, highlights this point. She complained that her father would not let her do anything because "he is afraid I will be like Americans." She protested that although—or perhaps because—her father had been in the Chicago area for sixteen years, he did not want her hanging around American girls because they "go out alone, talk back to their parents and don't help their mothers." She then reasoned that her father was right because he had seen how "American girls treat their mothers." On top of that, Zaragoza's father let her attend movies or social gatherings only two or three times a year when he came along as chaperone.[22]

Parents expected adolescent girls to be accompanied by either a parent or a sibling to social events involving adolescent boys or men. "I like to belong to the club because we get a chance to meet other girls and we have a lot of fun when we have our socials," stated a girl interviewed at Bird Memorial in the mid-1930s. She mentioned that they included boys but emphasized that only boys who were members of a club that met at Bird Memorial were invited. "On these occasions no parents are invited," she continued. "That's why we girls have so much fun. We don't have anybody watching us." These girls also resisted demands from parents that younger sisters participate in club meetings and parties. As in the case of women working outside of the home, more traditional parents "viewed with alarm" the fact that other neighborhood residents allowed unchaperoned girls to attend social events.[23]

Generational differences existed between older Mexican immigrants and their kids, whether the children were born in Mexico or the United States. While many men assimilated only to the extent necessary to get and keep a job, children were less worried about negotiating the cultural obstacles between being Mexican and being "American." When not using the facilities at a park or a community center, boys and girls organized their own recreational activities to emulate "what other American kids did." These activities included using cardboard or barrel sheaves to sled in the winter and broomsticks and smashed cans to play street hockey in the summer. Playing games with tops and marbles was also popular. Other activities for the boys included going to Eggers Grove, a recreational and picnic area on the east side of the South Chicago neighborhood which had wooded areas.

Anthony Romo recalled doing "whatever boys did in a field of strictly woods. Play around, Tarzan or Daniel Boone, or whatever. Then [we would] come home late at night." Swimming and fishing in Calumet Park or in a small lake behind U.S. Steel, going on bicycle trips, and making visits to a nearby dump to hunt for things of value were also common summer activities. The fact that first- and second-generation kids played previously unheard of games such as street hockey, Tarzan, and Daniel Boone underscores the generational rifts between children and parents in their attitudes towards acculturation into "American" society.[24]

Men and boys had many more opportunities to participate in team sports than did women and girls. The Chicago Park System, Bird Memorial, Common Ground, and the Catholic Youth Organization (CYO) at Our Lady of Guadalupe were the primary institutional organizers of youth baseball and basketball teams in South Chicago. Many youth first discovered Bird Memorial and Common Ground because of their baseball programs. Staff and volunteers at these organizations coached, officiated, reserved playing courts and fields for games, and organized tournaments where participants took home trophies. South Chicago's Mexican boys and girls also traveled to the Back of the Yards and the Near West Side neighborhoods to participate in

Fig. 8.2. Mayas youth baseball team, South Chicago, 1937. Photograph courtesy of the Southeast Chicago Historical Society (neg. no. 81-120-42t).

settlement house–sponsored activities, team sports, and tournaments. The settlement houses within these neighborhoods (the University of Chicago Settlement House serving the former and Hull House the latter) provided the large Mexican communities with group activities and leagues that in many cases were exclusively Mexican. Older boys who participated in baseball leagues also coached junior baseball teams. These teams, rarely lasting more than a single season, provided further interaction within the community as the junior teams played local and area teams, Mexican and non-Mexican. In addition to playing in baseball leagues organized by social service organizations and settlement houses, Mexican boys in South Chicago organized their own pick-up baseball games and softball leagues with teams named after the streets they lived on, such as the Baltimore Aces and the Buffalo Braves.[25]

Although little information exists about organized sports leagues for girls, 1934 Spanish-language newspaper articles mention the presence of Mexican girls' basketball teams. One article stated that "[t]he year of 1934 has unusually awakened the youth from both sexes to the basketball attraction," with the girls' teams "both being of first magnitude" and playing a "clean cut game" with a final score of 18 to 8. Although played at the University of Chicago Settlement House in the Back of the Yards neighborhood, the teams consisted entirely of *Mexicanas*, some of whom undoubtedly lived in South Chicago.[26]

Aside from the physical activity and sense of community built by the participants and spectators of the games, organized sports provided opportunities for Mexican South Chicagoans to become familiar with other neighborhoods and ethnic groups. Although Mexicans in South Chicago were not isolated from communities around them, local youth traveling to sporting events visited ethnic communities that they would otherwise avoid. One of the first Mexican boys' basketball teams, organized by a group of five friends and sponsored by Bird Memorial, was the Southern Arrows. Pete Martínez, an original member of the team, remembered "traveling everywhere." Through the Bird Memorial–sponsored basketball team, Martínez was exposed to a range of communities: "Oak Park, Villa Park, wherever they would tell us to go play, we would go play." José Cruz Díaz, another member of the team, remembered traveling all over Chicago playing basketball: "if we couldn't go [or] get somebody to take us, we used to take the streetcar to all these places." He added, "We went to some places that had never seen Mexicans at all whatsoever, and they used to ask us, 'Are you Mexican?' We said 'Yeah.' We were like an exhibit."[27]

The idea of being treated like an exhibit highlighted the perception that Mexicans were outsiders and unequal to members of the white and ethnic

European communities of Chicago. Despite this, Martínez credits the basketball team with making players aware "that there were other neighborhoods, other people, other groups." For many Mexican youth, their street had been their "whole world." As was the case for other ethnic youth who grew up in Chicago, travel outside the neighborhood and competition with other ethnic teams mitigated against the isolation of Mexican youth.[28]

The Catholic Church's ban on the use of Bird Memorial eventually forced the Southern Arrows to disassociate themselves from their home court. Claretian Father Anthony Catalina, who at the time was a priest at Saint Francis of Assisi, the Mexican Catholic Church in the city's Near West Side neighborhood, and would later be assigned to Our Lady of Guadalupe in South Chicago, confronted team leaders Pete Martínez and Angel Soto. In attempting to get members of the South Chicago community to use a Catholic facility that Catalina established on the Near West Side, the good father reminded the boys that Bird Memorial was not a Catholic Church. In response, they decided to cut the club's ties to Bird Memorial: "Instead of going against him, we'll just break up and do it on our own on the outside," said Martínez, "which we did." After parting ways, Martínez regretted the dissociation and the team disbanded a short time later. South Chicago community organizations continued to create youth baseball and basketball teams with the height of their popularity coming during the pinnacle of the depression.[29]

One indication of the popularity and importance of these activities to the community is that Spanish-language newspapers regularly covered the games and promoted team events. Thus, for example, a January 1932 article in *El Nacional* announced that the Atlas Sports Club started a new basketball team "headed by the young sportsman Mr. Angel Soto, who will act as manager, and captained by Augustin Mendoza." Despite the fact that this was a South Chicago team, the "debut" game was in the principal gymnasium at Hull House and was to "be attended by Mr. Rafael Aveleyra, Consul of Mexico, and the high functionaries of Hull House." The news coverage, the presence of the consul at the game, and the fact that the game was taking place in another neighborhood were all significant in ways that demonstrate the value the community placed on these games and on the interactions among the main Mexican communities.[30]

Within two months of its debut, the South Chicago Atlas basketball team had played seventeen games, winning fifteen of them. Atlas played teams from throughout the area. Their victories were against the Pirates of Indiana Harbor, the Silver Streaks, Cuauhtemoc Confederated Boys, Oak Park, Bethlehem, Park Manor, Villa Park, Maywood, Esses, and Chiefs.[31] The newspaper article celebrating this accomplishment reported that "The directors of the Theological Seminary of Chicago University have offered them an opportunity to show their mettle again by admitting them to a recently organized league which will compete for a silver trophy."[32] Playing against non-Mexican teams, and winning, provided the community with a nationalistic pride that reinforced the cultural "third-space" of Mexican South Chicago. The article continued by praising the team: "The Mexican Colony of South Chicago is gratified and quite satisfied with the accomplishments of this famous sport club which has given prestige to the name of Mexico." Linking the entire Mexican community of Chicago to this one South Chicago team, *El Nacional* went on to praise the team for all of the ways that it benefitted the community:

> The Atlas Sport Club is always making an effort to improve the team and broaden its activities. It now informs us of having planned a dance ... a ball which, without doubt will be attended by all of our young people. The place selected is the Community Hall of South Chicago, a place where the unit already referred to has given several social functions. The members of the Atlas will never cease their efforts to build up a team which will be the glory of Mexico and capable of facing any team from any other country. We congratulate these young men who have earnestly devoted themselves to the development of sports.[33]

Although the *Foreign Language Press Survey* translation of the article, the only extant remnant of the article, is a literal translation, the intent of the article is clear. The Atlas boys' basketball team, and the sports club that sponsored it, were important to the Mexican community of Chicago. In making the community look good, they were in the process of becoming "the glory of Mexico" and had already given "prestige to the name of Mexico." This linking of youth sports with the Mexican community's cultural nationalism is another indication of the level to which Mexicans in the Chicago area held pride in their *patria*.

Meanwhile, leaders of South Chicago's Mexican community created, advocated for, and managed other community-based organizations originally formed to support men's baseball and basketball teams. Baseball was the most popular team activity and spectator sport, peaking in appeal and participation during the depression and as a result of high unemployment and sporadic work. Because Mexican baseball teams were unable to use Bessemer Park until the late 1920s, earlier squads practiced and played in nearby open prairies. It was not be until 1929, when a flood of immigration came to the area, that the group made inroads to park access. It took the Great

Fig. 8.3. Yaquis baseball team, South Chicago, undated. Photograph courtesy of the Southeast Chicago Historical Society (neg. no. 81-120-42i).

Fig. 8.4. Mayas baseball team, South Chicago, undated. Photograph courtesy of the Southeast Chicago Historical Society (neg. no. 81-120-42d).

Depression and the large number of unemployed adults to open Bessemer Park to all Mexicans. By the turn of that decade, baseball games in Bessemer Park involving local teams regularly drew upwards of 1,000 people. Access to the other South Chicago public parks, Calumet and Russell Square, followed a similar timeline.[34]

The depression did not significantly improve recreational and leisure-time options for women. Although they participated by supporting the male athletes of the community through attendance at games and added domestic responsibilities, the vast majority of women did not have recreational opportunities despite the popularity of sports in the community.

The multi-dimensional sports-based organizations created to support teams eventually became critical cogs in the well-being of the community. Owner of one of the first Mexican grocery stores to serve the South Chicago Mexican community, Eduardo Peralta not only organized and managed a baseball team, but also recruited others to start their own teams. He founded the *Club Deportivo Yaquis* in 1932. According to one participant, the name was chosen after they "found out most of the players were from Sonora," a home of the Yaqui Indians. As chair of the *Club Deportivo Yaquis*, Peralta characterized his club in a 1936 WPA interview as having thirty-five members, who met once a month during the winter and more often during the summer. The organization eventually expanded to include basketball, football, and "indoor-ball" team.[35] Aside from the transparent benefits to the community of having a locally run team that brought people together for games, the importance of the organization to the community became more pronounced after the Yaquis sponsored fund-raising dances several times a year to take in money to buy uniforms and equipment.[36]

The Yaquis provided "no educational training, except that which is connected with our line of work." Although team members had been working at the steel mills before being laid off, the type of training mentioned by Peralta is open to speculation. Despite that fact that the Yaquis were not a mutual aid society that collected dues or provided guaranteed benefits in case of death or injury, team members helped other members when they were ill or were injured while playing for the team. Claiming that the organization's aim was "to do right in every way possible to the community, and to procure the welfare of our members at the same time," Peralta mentioned that the Yaquis had ties to the *Club Deportivo Monterrey* of South Chicago, also founded in 1932, and the *Sociedad Pro-Mexico*, an organization dedicated to promoting Mexican culture and patriotic holidays. The Yaquis were also members of the South Chicago Chamber of Commerce. Peralta significantly underestimated the overall significance of the club and its activities to the community in emphasizing only the benefits of their providing free entertainment and an economic boost by their purchasing team equipment within the barrio.[37]

By the mid-1930s, multipurpose clubs were more common. Although the *Club Deportivo Monterrey* remained a one-sport club (baseball), by 1936 it consisted of thirty members and held "social festivals" four times a year. Unlike the Yaquis, *Club Deportivo Monterrey* did serve as an actual mutual aid society. Its members paid ten cents a week to enroll in a disability plan that paid members two dollars and fifty cents a week for up to four weeks if unable to work because of sickness or accident.[38]

Like the Yaquis, the Pirates baseball team was an organization originally formed by members of the community. Started around 1927 by a group that included Gilbert Martínez and Angel Soto, the Pirates experienced firsthand the discrimination of other ethnic groups and park officials that limited their access to Bessemer Park.[39] These groups and officials forced the team to practice and play on sandlots and open prairies until team members discovered the baseball fields at Washington Park.[40]

Baseball became more popular with local Mexicans as the depression deepened. Members of the community were no longer working twelve to fourteen hour shifts, six days a week. As the number of people playing and watching baseball games grew, the Pirates merged with another team. Eustebio Torres, owner of a local newsstand and publisher of the short-lived *Anunciador* newspaper, named the new team the Excelsiors, after the major Mexico City newspaper that could be found on his newsstand. Team members speculated that Torres's motivation for choosing the name was a hope that the *Excelsior* newspaper would cover his team and pay more attention to the Mexicans of Chicago. As the larger organization that revolved around the baseball team took shape, the Excelsiors created a board of directors, president, vice president, and other officers, including non-playing members whose duties were to contribute "to the welfare of the team." Like the Yaquis and the Monterrey sports clubs, the Excelsior club held regular dances to raise funds. The organization eventually expanded to include an acting club that put on a production at the J. N. Thorpe School, a neighborhood public grade school. The play drew the attention, attendance, and support of members of the board of education as well as the Mexican consul. The Atlas and the Mayas, like the Excelsiors and the Yaquis, are teams with lasting legacies that remain part of the barrio's lore to this day.[41]

In "The Playing Fields of St. Louis," historian Gary Ross Mormino highlights the importance of sports (primarily soccer, football, baseball, and basketball), and organizations that contributed to the success of the teams, in the development and preservation of an Italian culture in St. Louis. Mormino, while focusing on the years between 1925 and U.S. entry into World War II, argues that "Sport encouraged not only the preservation of an ethnic subculture, but the preservation of the community itself."[42] Sports played a significant role in unifying a historically divided Italian immigrant community. In addition, he argues, the neighborhood athletic federation that emerged from the community's interest in sports became a "powerful signal of ethnic group identity."[43] A significant difference between Italians in St. Louis and Mexicans in South Chicago was the level of racial discrimination experienced by each group. Although working-class Italians did face discrimination in St. Louis, they had acquired a level of "whiteness" that remained elusive to Mexican immigrants in that city and in Chicago.

In *Making Lemonade out of Lemons,* José Alamillo argues that Mexicans in Corona, California, were able to use community baseball and sports clubs for a variety of purposes, including community formation. Unlike in South Chicago, the earliest Mexican teams and leagues in Corona were organized by large employers with the expressed goal of Americanizing and instilling discipline in Mexican workers. Within short order, *Mexicano* "sports clubs, mutual aid organizations, churches and small businesses" were creating teams and forming leagues. The most significant difference between these two communities is the in the evolution of these teams and organizations. In Corona, dynamic pre-existing organizations created baseball teams; in Chicago, support organizations were created either simultaneously or after the creation of the baseball teams.[44]

According to historian Douglas Monroy, *Mexicano* baseball in Los Angeles started as a "high society" sport as early as 1916. He argues that "Because of the high level of competition and the number of teams, a cross section of Mexican social classes undoubtedly took to the field together" in Los Angeles.[45] The first of these developments did not happen in Chicago possibly because of the tiny number of educated and "high society" Mexicans in the Second City and the newness of the Mexican migration to Chicago, especially in comparison to cities like Los Angeles that had once been part of Mexico. As for Monroy's second point regarding a "cross section of Mexican social classes"—there is no evidence that in Chicago baseball anyone outside of the working class "took the field." However, it is also important to keep in mind that Chicago's Mexican class structure was much more of a bottom-heavy triangle than in Los Angeles. One other interesting difference between Monroy's Los Angeles and Chicago or Corona involves the role of businesses. Unlike employer-initiated teams and leagues in Corona, teams in L.A. were organized and sponsored by local businesses. Although South Chicago Mexican business owners were involved in organizing teams and leagues, support for teams came primarily from the organizations created by team members and boosters; local businesses contributed to a lesser extent. Additionally, team names were not predominantly named after businesses while some neighborhood businesses were named after popular teams.[46]

As community leaders created more teams throughout Chicago's three major Mexican communities, as well as in Northwest Indiana, they formed leagues. The teams used at least one enclosed park, in Indiana Harbor, where lights were installed for night games and minimal entry fees could be charged to raise funds for the teams. This stadium had a capacity of 3,000. Game attendance by community members was impressive. José Cruz Díaz, who at the time was part of the Mayas baseball team, boasted about being a member of "one of the best baseball teams in South Chicago." During the height of the depression, the baseball team would draw upwards of 1,000 spectators in Bessemer Park as they played almost daily. "We had nothing else to do," he recalled. He reasoned that because of lack of jobs and "work was very slow," members of the community traveled throughout Chicago and Northwest Indiana by truck, streetcar or foot to watch the team play. According to a newspaper report, over 2,500 spectators witnessed the *Club Deportivo Monterrey* defeat the Drexel Square Athletic Club in Washington Park in September 1933. Because of the popularity and quality of the Excelsior and Atlas baseball teams, they traveled "over the middle west as semi-professional teams." Mexican-owned small businesses in South Chicago and the other area *barrios* named their businesses after popular local teams to capitalize on the teams' success and the high level of community pride.[47]

José Cruz Díaz found some positive that came out of the Great Depression. He credits it for strengthening the community through sports. Others felt that the sports teams and the organizations that sprang up to support teams and leagues did more than just boost morale and foster a sense of unity among Mexicans throughout Chicago and Northwest Indiana. They believed that the quality of the basketball and baseball teams in South Chicago earned them respect among some anti-Mexican elements in the Polish community. Sidney Levin, a classmate and friend of several of the Mexican athletes during this period, agreed that there was more respect for the ball players. Speaking of "Polish roughnecks," Levin explained that these roughnecks that "used to be out to get one or two lonely guys that used to walk to the game, stopped it completely." Levin argues that this was not "due to the respect or the fear of the numbers of the people who came to the game" but because of the quality of the games that were being played.[48]

Although the influence and respect that these organizations fostered because of the quality of the teams' athletic performances, the influence of organizations such as the Yaquis, the Excelsiors, the Atlas and the Monterrey extended far beyond the baseball field or the basketball court. The most visible non-sports activity sponsored by these organizations were the dances and concerts that organization leaders held to raise funds for expenses related to team equipment and activities. These dances and concerts widened the influence and importance of these organizations by providing venues for recreation and facilitated social gatherings that allowed community members, as well as some people from outside the community, to expand social networks, including inter-ethnic relationships. Although the organizations were not the only groups holding dances and concerts, they were usually the most widely anticipated and attended because of the popularity of the teams and their athletes' status in the community. Several venues existed for these dances. In the South Chicago area these included the basement of the Steel City Bank, Croatian Hall, the Masonic Temple, Columbus Hall, Lillian Hall, Eagle's Hall, and Lincoln Hall, not to mention "wherever they figured they could make the most money, or the place was available." South Chicago Mexicans also attended dances at Hull House's Bowen Hall. Organizers put on dances as part of festivals that included basketball tournaments involving teams from different Mexican communities. The musical groups that played at these dances included bands with the names "Los Cubanos" and "The Royal Castilians," both all-Mexican bands. As the organizations sponsored an increasing number of dances, the number of youth and adult bands quickly grew.[49]

An analysis of the dance announcement provides significant insight into 1939 South Chicago. This dance, celebrating the second anniversary of the Royal Castilians, was held on a Saturday night at the "well-known and elegant" Eagle's Hall. It included a jitterbug contest with prizes donated by a non-Mexican-owned jewelry and men's clothing store, and the evening included restaurant and bar services. Analyzing the display ads surrounding the announcement is just as valuable to get a feel for the importance of the event and the importance of *Mexicano* patrons to the businesses. The sponsors were mainly Mexican-owned businesses, but a few non-Mexicans interested in the community's business paid for an ad. The sponsors included a pool room, a couple of tailors, a barber shop, a lunch room, a tortilla factory that made home deliveries, a musical instrument shop, two grocery stores, and a restaurant. Within these businesses were two Indiana Harbor (East Chicago), Indiana businesses—the lunch counter and a garage. This indicates that significant connections and commuting existed between South Chicago and Indiana Harbor.

Clearly, the sports-centered organizations served several functions within the South Chicago's Mexican community. These groups facilitated social

SOUTH CHICAGO SERVICENTER
93rd South Chicago Ave.
Washing Lubrication
Mobilgas Mobil Oil

"EL NACIONAL"
Pool Room
Pase Ud. a Recrearse
FRANCISCO MARTINEZ, Prop.
(EL GRANDE)
8838 Buffalo Ave. So. Chicago, Ill.

LA RIVERA
SASTRERIA
Aseo y Planchado de Trajes
de Damas y Caballeros
Hacemos Trajes a la Medida
Entregamos a Domicilio
F. CABALLERO,
Tel. So. Chicago 1264
9056 Brandon Ave. So. Chicago

LA PURISIMA
Tienda de Abarrotes y Carniceria
Panaderia y Fabrica de Tortillas
Tel. So. Chicago 0953
3240 E. 91st St.
CELEDONIO CASTILLO, Prop.

GRAN BAILE
2do. ANIVERSARIO
De La Orquesta
Royal Castilians
SAB. 13 DE MAYO - 1939
A Las 8: p. m.
- En El Conocido y Elegante -
EAGLE'S HALL
9231 Houston Ave.
PREMIOS DONADOS POR:
Cole & Young, Conocidos Joyeros y
Weiss & Weiss, Especialistas en Ropa para Caballeros.
JITTER BUG CONTEST
Premios En Efectivo
ADMISION:
Caballeros 40c Damas 25c
Servicio de Cantina y Restaurant

SILOS BARBER SHOP
Hat Cleaning and Blocking
F. SILOS, Prop.
Phone So. Chicago 2835
3211 E. 90th St.

EL INDIO
SASTRERIA
Limpiamos, Planchamos y
Reparaciones De Ropa
PAUL ARELLANO, Prop.
Tel. So. Chicago 2835
3209 E. 90th St.

Tony Polarek
Bismarck Beer
So. Chicago 1169
8901 Houston Ave.

TOLTEC
RESTAURANT
Ofrece a Ud. Esmerado Servicio y
Limpieza. Pase y se Convenzera
M. HERNANDEZ, PROP.
9006 Buffalo Ave.
TEL. SAGINAW 3784
SO. CHICAGO, ILL.

Gonzalez y Vega
GARAGE
COMPOSTURAS GARANTIZADAS
3854 Penn. Ave. Indiana Harbor, Ind.

"La Gardenia"
LUNCH ROOM
Le Ofrece a Ud. un Servicio Esmerado y Comodo
Cuartos con o sin Asistencia
477 Penn. Ave. Indiana Harbor Sra. Reynalda P. Robles, Prop.

Compliments of
Wm. A. ROWAN

CORDERO'S
Universal Radio Service
Para Localizar el Defecto en su Aparato Radio-Receptor usamos el "Rayo X del Radio", Lo mas Moderno de nuestro Equipo, se requiere menos tiempo en el Analisis de su aparato y por lo consiguiente el precio de reparacion es menos.
Regent 3628 3315 E. 89th St.

Carl Fischer Inc.
306 So. Wabash Ave.
Chicago, Ill.
Everything in Music &
Musical Instruments

EL
AZTECA
Tortillas y Masa
Entregos a Domicilio
P. CHAVEZ
Saginaw 0791 3243 E. 91st St.

Cortesia
-De-
United Tea Stores
La Tienda
Colorada

Fig. 8.5. Announcement for May 13, 1939, dance in Eagle's Hall. Photograph courtesy of the Southeast Chicago Historical Society (neg. no. 81-120-44).

Fig. 8.6. Justino Cordero in his South Chicago radio shop. Photograph courtesy of the Southeast Chicago Historical Society (neg. no. 81-120-15).

activities for adults and youth and provided opportunities for better communication and fellowship within and among the Mexican clusters of the Chicago area. Because of the respected status of the teams and athletes in the community, these organizations quickly—and successfully—expanded to provide other services and diversions for the community during time of crisis.

As a result of their need to expand the space available to them for leisure-time activities during the interwar years, Mexicans in South Chicago organized and rallied around organizations that helped open recreational spaces and opportunities despite the fact that ethnic prejudice, gender inequalities, and lack of resources created physical, economic, and psychological boundaries. Leaders emerged and used their sports teams and organizations as vehicles to push beyond their externally imposed boundaries and claim their rights as residents of this country. Organizations like the South Chicago Community Center and Common Ground were important, but members of the community also took the initiative to push the externally imposed boundaries and establish teams. The teams evolved into multi-functional organizations that proved to be anchors for many Mexicans who struggled financially and psychologically to remain and persevere during the Great Depression and the repatriation movement. Consequently, these clubs, their organizers, and their participants improved their community's quality of life during and long after the Great Depression. They also represented yet another way Mexicans in the steel barrio of South Chicago transcended externally imposed boundaries to develop a strong and long-lasting Mexican community in South Chicago.

EPILOGUE

Why should we care about the creation and development of a distinct Mexican community in South Chicago? Neither baseball nor basketball nor the consul nor Mercedes Rios single-handedly guided Mexicans in South Chicago through the community's early years and through the Great Depression. These individual and community histories—the stories of people, organizations, and their physical surroundings—shed light on *Mexicano* life in a place at once far from the border and within the industrial heart of the United States. Ninety-five years after the first wave of Mexican immigrants came to Chicago to work the railroads, Mexican Chicagoans have developed into a major political, economic, cultural, and social force. The rally on March 10, 2006, put on display the energy, political will, and organization of a Mexican community that scholars and national activists have often overlooked.

South Chicago Mexicans created a cultural, intellectual, and political community that prepared the group, albeit a smaller one, to survive the Great Depression. They laid the foundations for a group that survived neglect by city leaders and the closing of the steel mills. Through adaptation of culture, through internal and external negotiation on the use of English, and through the creation of recreational, social, and mutual aid organizations, Mexicans adapted their environment in order to persist and create a strong community that survived economic hardship and discrimination.

Mexican repatriation at the national and local level profoundly affected the depression-era Mexican experience in the United States. Although forced repatriation in Chicago never approached the numbers in California or the percentages of total Mexican residents in Detroit or Northwest Indiana, the movement of Mexicans out of South Chicago caused substantial changes in the community. More importantly for many *Mexicanos*, the threat of repatriation always hovered, creating a sense of uncertainty and insecurity beyond the economic crises of the depression.

Historians led by Vicki Ruiz and George Sánchez argue that the period surrounding the Great Depression served as a critical one in the formation of a Mexican-American identity in the United States. In South Chicago, resistance to discrimination, harassment, and pressures to "Americanize" brought together Mexicans from within and immediately around the neighborhood to form a South Chicago Mexican community. The pre-1940 South Chicago Mexican community played an integral part in the development of a greater, Chicago-area Mexican community. A common, yet fluid identity helped the community survive in lean years and thrive in prosperous years. Historians Lilia Fernández and Mike Amezcua both demonstrate the overall vitality and complexity of *Mexicano* and trans-Latino communities in post–World War II Chicago.[1] As American-born *Mexicanos* continued to live in a post-steel South Chicago and Latin American immigrants continue to see opportunities in the area, the community's influence grew and still does. In 2010, Chicago's Latino population was the fifth largest in the country, with almost 800,000 people of Latin American descent, making up almost 30 percent of the city's population.

The history of Mexican South Chicago matters. It matters because it illustrates how resistance and organization can help an immigrant community grow into a vibrant and powerful part of the larger local culture and society. Mexicans in South Chicago countered efforts by Americanization proponents to eliminate Mexican cultural practices and celebrations. They learned English only to the extent necessary to advance economically and created organizations that supported public displays of cultural unity. Mexicans negotiated within Americanization programs by learning what they felt they needed to survive and persist while being cognizant of the fact that many assimilationists expected complete Americanization. What reformers saw as their failure to take advantage of the resources offered or to enact the lessons supposedly learned about how to be "American" were in reality deliberate acts of resistance by Mexicans in South Chicago.

By recognizing the significance of the Mexican sojourner attitude, public displays through sports or celebrations, and the negotiation of the use of English, we can begin to delve into the strategies that helped Mexicans in South Chicago cope with the oppressive environment that surrounded them. Many Mexican immigrants to South Chicago planned to return to Mexico and did not actively seek to Americanize. Only a few sought citizenship. Mexicans who took advantage of English classes in order to improve employment prospects still saw Spanish as a critical component of their cultural and political identity. Many returned to Mexico voluntarily or were forcibly repatriated during the Great Depression. Others who never returned to Mexico nevertheless kept alive the dream of someday returning. The longer they stayed in South Chicago, the more entrenched they became in the social and cultural life of their local community. These ties bound them to their community in South Chicago, further strengthening the Mexican community that lived in the shadow of steel.

I write this last part of the book in the summer of 2012. The economic downturn is in its fourth year, and high unemployment has given energy to nativists and others who loudly distinguish between the white American "us" and the racialized immigrant "them." The significance of the creation, development, and cultural survival of the Steel Barrio transcends both space and time. The histories of Mexicans migrating to South Chicago—coping with racial discrimination and harassment, adapting to their environment while changing the environment around them, knowing when to leave and when to stay, or just doing what it took to survive—are as relevant today as they were eighty years ago.

States throughout the American South and West have proposed and passed laws that are aimed at discriminating against and harassing anyone who might look like a Latin American or Asian immigrant. As it was intended to do, this legislation is making life difficult for those of Latin American descent regardless of citizenship or immigration status. Learning lessons from the experiences of the *Mexicano* community and the actions of "white" society during the Great Depression can help immigrants and advocates find ways to redirect the national immigration debate away from the current "us" versus "them" rhetoric towards a constructive dialogue about the role of immigrants in American society today. The early twenty-first-century United States looks much different from that of the 1920s and 1930s, but similarities abound with regard to immigrant and popular reactions to economic crisis in both eras. The recent anti-immigrant efforts parallel many of those of nativists during depression-era America. Like Chicago's Mexican immigrants of the 1930s, today's Latino and Latina immigrants to Alabama and the American South are in the early years of settlement and community formation. Like the Chicago Mexican immigrant community of 1930, that of Alabama in 2011 is less than two decades old.

The organizing and resistance against today's anti-immigrant laws is breathtaking. Mexican community leaders, Mexican diplomats, social workers, and organizations like the Immigrant Protective League (IPL) were critically important actors in supporting the Mexican community during the Great Depression. Today, it is community leaders, religious leaders, advocates, community organizers, and the new and unexpected civil-rights coalitions that are forming to fight anti-immigrant waves throughout the country. Coalitions between organizations that are historically linked to fighting for African-American justice and those that have traditionally focused on non-immigrant Latino/as have joined with immigrant rights groups and religious organizations to create an unlikely, but strong, coalition.

One particular event in Birmingham, Alabama, dramatically illustrates the coalition's influence and potential. On November 22, 2011, I stood outside Birmingham's 16th Street Baptist Church with about 2,200 other people who arrived too late to find a seat inside the 2,000-seat church. This was the very site where four African-American girls were murdered by a Klu Klux Klan bomb on Sunday morning, September 15, 1963. This 2011 event was more then a pep rally or a protest. About 4,200 men, women, and children had come to this hallowed ground of the African-American civil rights movement to celebrate the launch of an immigrant civil rights campaign called "One Family/One Alabama." We watched on a jumbo screen outside the church as African-American, Latino, and Anglo-American speakers spoke about the need to organize and protest against de jure and de facto discrimination and harassment in Alabama, a state that lives in the shadow of its past. One after the other, leaders of Birmingham's civil rights movement of the 1960s invoked the spirit of the four little girls. That tragic event became a national catalyst for the pursuit of equality for African Americans then. Now, African-American community leaders had suggested the use of the 16th Street Baptist Church for this rally to emphasize the importance of African Americans joining with Latinos in fighting against Alabama's new anti-immigrant law and all other discrimination. Speaking from the pulpit to the crowd inside and outside of the church, retired Chief Federal Judge U. W. Clemons, a civil rights activist and Alabama's first African-American federal judge, called for unity as he repeated the famous slogan "an injury to one is an injury to all." He told Birmingham's veteran civil rights activists, the "foot soldiers" of 1960s Birmingham, to put their "marching shoes back on" and fight back against the oppression of immigrants in Alabama. The excitement in the multicultural crowd was palpable as cheers and tears filled the sanctuary and the street. On that night, it seemed as if Alabama's governor and legislative leaders had awoken a sleeping giant.

The politics of hate created an impressive alliance and an energy that not only linked people, but organizations. The National Association for the Advancement of Colored People (NAACP), the League of United Latin American Citizens (LULAC), the United Farm Workers of America (UFW), the Service Employees International Union (SEIU), the Greater Birmingham

Ministries, the Southern Poverty Law Center, the Hispanic Interest Coalition of Alabama, the Alabama Coalition for Immigrant Justice, Alabama Appleseed, the Birmingham City Government, and other organizations joined to vocally oppose the wave of anti-immigrant rhetoric and legislation in Alabama and the rest of the American South.

Today's struggle in Alabama has an unlikely link to Latino Chicago. On the afternoon before the rally, Representative Luis V. Gutiérrez led members of the U.S. Congress to Birmingham for a congressional ad-hoc hearing on the impact of Alabama's anti-immigrant laws. An unusually high number of members of Congress—eleven Democratic members—attended the hearing held at the chambers of the Birmingham City Council. Congressman Gutiérrez, considered the highest-ranking Latino in federal elected office, is of Puerto Rican descent and represents the largest Puerto Rican and Mexican communities in Chicago. Members of Alabama's immigrant community see Gutiérrez as an advocate and leader in the fight against nativism and discrimination in the state.

American history is rife with examples of increasing anti-immigrant nativism and nationalism as responses to a faltering economy and rising unemployment. Governments and civic organizations promote deportation as a solution to the high unemployment rate and the failing economy. During the Great Depression and during the recession that started in 2008, Americans turned their frustrations and anger towards Mexican and other Latin American immigrants, regardless of immigration status. The catchword in the 1930s was "repatriation." The catchwords today are "unlawful presence," "deportation," and "self-deportation." Southern states created a patchwork of laws codifying discrimination and harassment that aimed to, and at times succeeded in, making immigrant life intolerable for most and impossible for many. In the early 1930s, the Great Depression and repatriation hit a new South Chicago Mexican community that was just over ten years old. At the time of Alabama's economic downturn in 2010, the majority of people in Alabama's Mexican and Guatemalan immigrant communities had been in the state for fewer than a decade. The pressure to "Americanize" these non-white immigrants was high in interwar South Chicago. Much of the language used by advocates and nativists in the early twenty-first century seemed to come from a lack of historical awareness. Similar language was popular during the Great Depression as repatriation campaigns and mass deportation round-ups united nativist groups, businesses, and governments to rid Western and Midwestern cities of brown-skinned "foreigners" who "took our jobs." The fact that Mexican immigrants were recruited and welcomed back to the factories and the fields when World War II caused a labor shortage seems to be lost on nativists today. The 2011 harvests in Georgia and Alabama were busts as produce rotted in the fields shortly after those states passed anti-immigration legislation. In 2012, many farmers in both states drastically reduced the volume of their produce because of labor shortage concerns.

As the Mexican community in South Chicago grew, members changed their surroundings to provide an accepting and welcoming environment for those in the community. As small town Alabama and Georgia grew with the entry of Mexican and Guatemalan immigrants, many white locals became increasingly alarmed over their changing communities. Spanish-language signs on Main Street and the sound of Spanish conversations throughout town threatened locals who preferred the nostalgia of a quiet, all-native Southern town to a growing, multi-cultural American village. As in depression-era South Chicago, locals blame rising unemployment on "Mexicans" instead of larger economic trends.

Learning the stories of how Mexican South Chicago survived and became part of a vibrant twenty-first-century Latino Chicago can help scholars, organizers, and advocates of Latinos in the American South today. In addition to the racialization of Mexicans as less than white, Mexican immigrants and nativist locals fight the same battles over the sojourner attitude, cultural celebrations, and language use in 2010s Alabama as they did in 1930s Chicago. The stakes may be higher today as South Chicago's police car number four is now any Alabama state or local patrol car and many Judge Greens dot the Alabama landscape. Immigrant advocates in Alabama are fighting institutionalized racism not present in Chicago even at the peak of the Great Depression. Yet immigrant advocates and leaders of today's civil rights movements can benefit by studying and beginning to understand each other's journeys, adventures, and histories to fully appreciate the new alliances and the potential they hold to change the Southern landscape of tomorrow.

NOTES

INTRODUCTION

1. For further clarification, see the note on terminology and labels at the end of the introduction.
2. Amalia Pallares and Nilda Flores-González, eds., *¡Marcha!: Latino Chicago and the Immigrant Rights Movement* (Urbana: University of Illinois Press,2010), xi.
3. Xóchitl Bada, "Mexican Hometown Associations in Chicago: The Newest Agents of Civic Participation," in *¡Marcha!*, 146.
4. Ibid.
5. Scott Fornek, "Chicago 'Giant' Put Rest of Country on Notice," *Chicago Sun-Times*, Chicago, April 2, 2006; Anna Johnson, "Activists Say It's No Surprise Chicago Immigration Rally Set Tone for the Nation," *Associated Press*, April 16, 2006; "Key Dates in Immigration Protest Movement," *Associated Press*, April 12, 2006.
6. U.S. Census Bureau, "Hispanic or Latino by Type, Chicago City, Illinois," Summary File 1, Table PCT 11, *2010 Census*. http://factfinder2.census.gov/faces/ tableservices/jsf/pages/productview.xhtml?pid=DEC_10_SF1_QTP10&prodType= table.
7. Gilbert Martínez, "Interview by Jesse J. Escalante," in *Jesse Escalante Oral Histories, Global Communities Collection*, Chicago History Museum. Escalante, the interviewer, grew up in the community, was a de facto community leader and a supervisor at a government office. Going back to school to earn a graduate degree, he conducted interviews like this one with Gilbert Martínez as part of his research.
8. A few representative examples of this vast literature include: Gabriela F. Arredondo, *Mexican Chicago: Race, Identity, and Nation, 1916–39* (Urbana: University of Illinois Press, 2008); James R. Barrett, *Work and Community in the Jungle: Chicago's Packinghouse Workers, 1894–1922* (Urbana: University of Illinois Press, 1987); Lizabeth Cohen, *Making a New Deal: Industrial Workers in Chicago, 1919–1939* (New York: Cambridge University Press, 1990); Lilia Fernández, *Brown in the Windy City: Mexicans and Puerto Ricans in Postwar Chicago* (Chicago: University of Chicago Press, 2012); James R. Grossman, *Land of Hope: Chicago, Black Southerners, and the Great Migration* (Chicago: University of Chicago Press, 1989); Thomas A. Guglielmo, *White on Arrival: Italians, Race, Color, and Power in Chicago, 1890–1945* (Oxford: Oxford University Press, 2003); Thomas J. Joblonsky, *Pride in the Jungle: Community and Everyday Life in Back of the Yards* (Baltimore: Johns Hopkins University Press, 1993); James B. LaGrand, *Indian Metropolis: Native Americans in Chicago, 1945–1975* (Urbana: University of Illinois Press, 2002); and Dominic A. Pacyga, *Polish*

Immigrants and Industrial Chicago: Workers on the South Side, 1880–1922 (Columbus: Ohio State University Press, 1991).

9. For a look at Gamio's interviews throughout the United States, see *El Inmigrante Mexicano: La Historia De Su Vida: Entrevistas Completas, 1926–1927* (México: University of California CIESAS, 2002).
10. Paul Schuster Taylor, *Mexican Labor in the United States: Chicago and the Calumet Region, University of California Publications in Economics*, vol. 7, no. 2 (Berkeley: University of California Press, 1932); Manuel Gamio, *Mexican Immigration to the United States; A Study of Human Migration and Adjustment* (Chicago: University of Chicago Press, 1930); and Manuel Gamio, *The Mexican Immigrant: His Life-Story* (Chicago: University of Chicago Press, 1931).
11. While the vast majority of graduate research produced during this period is of little value to the study of the Mexican community of South Chicago, two clear exceptions are Edward Jackson Baur's work on "Delinquency Among Mexican Boys in South Chicago" (M.A. thesis, University of Chicago, 1938), and Anita Edgar Jones's work on "Conditions Surrounding Mexicans in Chicago" (M.A. thesis, University of Chicago, 1928). Baur, drawing from a seemingly rich set of interviews conducted as part of a WPA study of over 2,000 South Chicago Mexican families (data that no longer survives), provides examples of immigrant travel experiences and of family life in South Chicago. Jones presents vivid examples of housing and neighborhood conditions for Mexicans in various Chicago locales. Also useful in the discussion of Mexican South Chicago is Eunice Felter's work on "The Social Adaptations of the Mexican Churches in the Chicago Area" (M.A. thesis, University of Chicago, 1941). Felter provides a limited, but useful "snapshot" of various Mexican churches throughout Chicago just before U.S. entry into World War II, albeit with little analysis of the development of these churches over time. A scattering of articles and WPAreports appeared in the late 1930s and 1940s, including: Ruth S. Camblon, "Mexicans in Chicago," *The Family* 7, no. 7 (1926); Norman D. Humphrey, "The Detroit Mexican Immigrant and Naturalization," *Social Forces* 22, no. 3 (1944); Norman D. Humphrey, "The Migration and Settlement of Detroit Mexicans," *Economic Geography* 19, no. 4 (1943); and Robert C. Jones and Louis R. Wilson, *The Mexican in Chicago, The Racial and Nationality Groups of Chicago: Their Religious Faiths and Conditions* (Chicago: Comity Commission of the Chicago Church Federation, 1931). Both Camblon and Humphrey provide paternalistic overviews of Mexicans in their respective cities, many times highlighting perceived problems caused by Mexicans.
12. Studies of Mexicans in the Midwest include Juan R. García, *Mexicans in the Midwest, 1900–1932* (Tucson: University of Arizona Press, 1996); James B. Lane and Edward J. Escobar, *Forging a Community: The Latino Experience in Northwest Indiana, 1919–1975* (Chicago: Cattails Press, 1987); F. Arturo Rosales, *Pobre Raza! Violence, Justice and Mobilization among Mexico Lindo Immigrants, 1900–1936* (Austin: University of Texas Press, 1999); Francisco Arturo Rosales, "Mexican Immigration to the Urban Midwest During the 1920s" (Ph.D. diss., Indiana University, 1978); Ciro Sepúlveda, "La Colonia Del Harbor: A History of Mexicanos in East Chicago, Indiana, 1919–1932" (Ph.D. diss., Notre Dame University, 1976); Dennis Nodín Valdés, *Al Norte:: Agricultural Workers in the Great Lakes Region, 1917–1970* (Austin: University of Texas Press, 1991); Dionicio Nodín Valdés, *Barrios Norteños: St. Paul and Midwestern Mexican Communities in the Twentieth Century* (Austin: University of Texas Press, 2000); Dionicio Nodín Valdés, *El Pueblo Mexicano en Detroit y*

Michigan: A Social History (Detroit: Wayne State University Press, 1982); Zaragosa Vargas, *Proletarians of the North: A History of Mexican Industrial Workers in Detroit and the Midwest, 1917–1933* (Berkeley: University of California Press, 1993).

13. Louise Año Nuevo Kerr, "The Chicano Experience in Chicago: 1920–1970" (Ph.D. diss., University of Illinois at Chicago Circle, 1976).
14. Ibid., 14.
15. Arredondo, *Mexican Chicago*.
16. John Henry Flores, "On the Wings of the Revolution: Transnational Politics and the Making of Mexican American Identities" (Ph.D. diss., University of Illinois at Chicago, 2009); Mike Amezcua, "The Second City Anew: Mexicans, Urban Culture, and Migration in the Transformation of Chicago, 1940–1965" (Ph.D. diss., Yale University, 2011).
17. For more on the importance of the 1930s in the development of a new Mexican-American identity see George J. Sánchez, *Becoming Mexican American: Ethnicity, Culture, and Identity in Chicano Los Angeles, 1900–1945* (New York: Oxford University Press, 1993), 11–13. George Lipsitz, in *Time Passages: Collective Memory and American Popular Culture* (Minneapolis: University of Minnesota Press, 1990), 16, argues that culture creates conditions that allow for the expansion or improvement of everyday life "by informing [the present] with memories of the past and hopes of the future; but they also engender accommodation with prevailing power realities." It is through this new culture that Mexicans in South Chicago accommodated some demands from the dominant culture while not giving up the parts of their common Mexicanness that served as unifiers of the community in hopes for a better, post-depression life.
18. Camille Guerin-González, *Mexican Workers and American Dreams: Immigration, Repatriation, and California Farm Labor, 1900–1939* (New Brunswick: Rutgers University Press, 1994), 2; Sánchez, *Becoming Mexican American*, 224–25; Vargas, *Proletarians of the North*, 176–90.
19. Año Nuevo Kerr argues that the majority of Mexicans who stayed in Chicago were in stable family units ("Chicano Experience," 72–73). This argument implies that those in family units were much less willing to disrupt family life and schooling of children to return to an uncertain future in Mexico.
20. Arnoldo de León, *Ethnicity in the Sunbelt: A History of Mexican Americans in Houston* (Houston: Mexican American Studies Program, University of Houston, 1989); Richard A. García, *Rise of the Mexican American Middle Class* (College Station: Texas A&M University Press, 1991); Thomas E. Sheridan, *Los Tucsonenses: The Mexican Community in Tucson, 1854–1941* (Tucson: University of Arizona Press, 1986).
21. This group included diplomats at the Mexican consulate in Chicago, academics at the University of Chicago, medical professionals, and senior corporate officials working within international divisions of Chicago-based corporations. These elite Mexicans had, for the most part, lighter complexions and sought to maintain an identity separate from working-class Mexicans.
22. Valdés, *Barrios Norteños*, 103–104.
23. Sarah Deutsch, *No Separate Refuge: Culture, Class and Gender on an Anglo-Hispanic Frontier in the American Southwest, 1880–1940* (Oxford: Oxford University Press, 1987), 5.
24. Vicki L. Ruiz, *Cannery Women, Cannery Lives: Mexican Women, Unionization and the California Food Processing Industry, 1930–1950* (University of New Mexico Press, 1987); Julia Kirk Blackwelder, *Women of the Depression: Caste and Culture in San Antonio, 1929–1939* (College Station: Texas A&M University Press, 1984).

25. Donna Gabaccia, *From the Other Side: Women, Gender, and Immigrant Life in the U.S., 1820–1990* (Bloomington: Indiana University Press, 1995).
26. Vicki L. Ruiz, *From Out of the Shadows: Mexican Women in Twentieth-Century America* (Oxford: Oxford University Press, 1998).
27. For more on the concept of community and "sense of community," see Jonathan D. Amith, "Place Making and Place Breaking: Migration and the Development Cycle of Community in Colonial Mexico," *American Ethnologist* 32, no. 1 (2005); Thomas Bender, *Community and Social Change in America* (New Brunswick: Rutgers University Press, 1978); Edward J. Escobar, "The Forging of a Community," in Lane and Escobar, *Forging a Community*; García, Guiuliani, and Wiesenfeld, "Community and Sense of Community"; Douglas A. Hurt, "Defining American Homelands: A Creek Nation Example, 1828–1907," *Journal of Cultural Geography* 21, no. 1 (Fall/Winter 2003). García et al. argue that "As time goes by, neighbors' interactions and mutual understanding can become more intense because they have the same needs and common problems" (730). Hurt, in his discussion of American-Indian homelands, introduces several concepts useful for conceptualizing community, including the notion that "this subjective [group] identity is reinforced by regular social interactions (such as ceremonies, religious activities, and festivals) that is further enhanced by the clustered nature of the group. While residents of the homeland had differing personal views, cultural attitudes, and social affiliations, members identify themselves as belonging to a specific distinct group" (21–22). In his study of migration and community in a Mexican colonial village, Amith argues that among migrants, "old communities were continually reproduced through the regenerative village practices of indigenous peasants, who were in this manner linked to spaces pregnant with historical memory and communal identity" (162).
28. David G. Gutierrez, "Migration, Emergent Ethnicity, and the 'Third Space': The Shifting Politics of Nationalism in Greater Mexico," *Journal of American History* 86, no. 2 (1999): 488.
29. Scholars generally agree that the Great Migration from Mexico occurred from the 1880s, when the railroad entered the interior of Mexico, until the Great Depression. For a discussion of the debate surrounding the factors causing the Great Migration, see Douglas Monroy, *Rebirth: Mexican Los Angeles from the Great Migration to the Great Depression* (Berkeley: University of California Press, 1999), 75–106.
30. Monroy, *Rebirth*, 4.
31. Renato Rosaldo, "Assimilation Revisited," in *In Times of Challenge: Chicanos and Chicanas in American Society*, ed. Juan R. García, Julia Currey Rodriguez, and Clara Lomas (Houston: Mexican Studies Program, University of Houston, 1988), 44.
32. Richard Griswold del Castillo, *The Los Angeles Barrio, 1850–1890: A Social History* (Berkeley: University of California Press, 1979); Sánchez, *Becoming Mexican American*, 6.
33. De León, *Ethnicity in the Sunbelt*, 115.
34. Sánchez, *Becoming Mexican American*, 13.

CHAPTER 1

1. Gilbert Martínez, in "Interview by Jesse J. Escalante," in *Jesse Escalante Oral Histories, Global Communities Collection*, Chicago History Museum; "Big Battle Near Torreon: Villa Leads Army Against 5,000 Carranza Men," *New York Times*, December 17, 1914; "Battle of Torreon: Won By Mexican Rebels," *The Advertiser* (Adelaide, South Australia), April 4, 1914. For more on early border enforcement and the ease with which

Mexicans crossed legally into the United States in the early twentieth century, see chapter 5 of Patrick W. Ettinger, *Imaginary Lines: Border Enforcement and the Origins of Undocumented, 1882–1930* (Austin: University of Texas Press, 2009).

2. Martínez, in "Interview by Jesse J. Escalante," *Jesse Escalante Oral Histories, Global Communities Collection*, Chicago History Museum.
3. I will provide other specific examples of the migrant journey to Chicago in the first two chapters of this book. For further individual examples, see Gabriela F. Arredondo, *Mexican Chicago: Race, Identity, and Nation, 1916–39* (Urbana: University of Illinois Press, 2008), 18–20.
4. Ibid., 20. Gabriela Arredondo emphasizes the "middling and propertied peoples migrate more easily" because of the "mobility made possible by privilege." Although middling and propertied peoples did move more easily, most who identified as working class once in the United States did so in the same manner as Gilbert Martínez's family.
5. For more on the historical use of *enganchistas* by American farmers and railroad companies, see Francisco E. Balderrama and Raymond Rodriguez, *Decade of Betrayal: Mexican Repatriation in the 1930s*, revised ed. (Albuquerque: University of New Mexico Press, 2006).
6. Lacy Simms interview by Paul S. Taylor, box 11, file 32, *Paul Schuster Taylor Papers*, BANC MSS 84/38 c, The Bancroft Library, University of California, Berkeley.
7. Anita Edgar Jones, "Conditions Surrounding Mexicans in Chicago" (M.A. thesis, University of Chicago, 1928), 51.
8. E. W. Burgess and Charles Shelton Newcomb, *Census Data of the City of Chicago, 1920* (Chicago: University of Chicago Press, 1931); Immigrant Protective League, "Report on Mexicans for Hull House Yearbook, 1927," in *Immigrant Protective League Papers*, Special Collections, University of Illinois at Chicago, Chicago.
9. Immigrant Protective League, "Mexican Report for Hull House."
10. E. W. Burgess, Charles Shelton Newcomb, and University of Chicago. Social Science Research Committee, *Census Data of the City of Chicago, 1930* (Chicago: University of Chicago Press, 1933), 33–34.
11. The standard social history on the impact of World War I in the United States is David M. Kennedy, *Over Here: The First World War and American Society* (New York: Oxford University Press, 1980). Stéphane Audoin-Rouzeau and Annette Becker mix accounts of World War I battles with a study of the effects of violence on soldiers and civilians in Europe in their *14-18, Understanding the Great War* (New York: Hill and Wang, 2002). Despite the fact that the United States government started compiling statistics on Mexican immigration to the United States in 1908, official immigration numbers are extremely unreliable because of the ability of Mexican immigrants to bypass official ports of entry when entering the United States; Dillingham Commission, *Reports of the Immigration Commission, Dictionary of Races and Peoples* (Washington, DC: Government Printing Office, 1911), 96. The fact that many Mexican immigrants in the United States lacked legal status undoubtedly contributed to a significant undercounting of Mexicans by census takers as many Mexicans avoided contact with anyone considered a government official.
12. Manuel Gamio, *Mexican Immigration to the United States; A Study of Human Migration and Adjustment*, (Chicago: University of Chicago Press, 1930), 47.
13. Jeffrey Marcos Garcilazo, "'Traqueros': Mexican Railroad Workers in the United States, 1870–1930" (Ph.D. diss., University of California, Santa Barbara, 1995), 68; Mark Reisler, *By the Sweat of Their Brow: Mexican Immigrant Labor in the United States, 1900–1940.*

(Westport, CT: Greenwood, Press, 1976), 14–17. Vicki Ruíz, *From Out of the Shadows: Mexican Women in Twentieth-Century America* (New York: Oxford University Press, 1998), 7.

14. Reisler, *By the Sweat of Their Brow*, 14–17. For a detailed explanation of the Mexican Revolution, see Héctor Aguilar Camín and Lorenzo Meyer, *In the Shadow of the Mexican Revolution: Contemporary Mexican History, 1910–1989*, Translations from Latin America Series (Austin: University of Texas Press, 1993), 1–158. The standard English-language work for the Mexican Revolution is Alan Knight, *The Mexican Revolution*, 2 vols., *Cambridge Latin American studies 54* (New York: Cambridge University Press, 1986). Newer work includes William H. Beezley and Colin M. MacLachlan, *Mexicans in Revolution, 1910–1946: An Introduction* (Lincoln: University of Nebraska Press, 2009); Colin Michael Gonzales, *The Mexican Revolution, 1910–1940* (Albuquerque: University of New Mexico Press, 2002); and Friedrich Katz, *The Life and Times of Pancho Villa* (Stanford: Stanford University Press, 1998).
15. Aguilar Camín and Meyer, *In the Shadow of the Mexican Revolution: Contemporary Mexican History, 1910–1989*, 84–88. For the Cristero Rebellion, see: Jean A. Meyer, *The Cristero Rebellion: The Mexican People between Church and State, 1926–1929* (Cambridge: Cambridge University Press, 2008); Christopher Boyer, *Becoming Campesinos: Politics, Identity, and Agrarian Struggle in Postrevolutionary Michoacán, 1920–1935* (Stanford: Stanford University Press, 2003), chapter 5; Jean Meyer and Enrique Krauze, *La Cristiada. La Guerra de los Cristeros*, 3 vols. (México City: El Colégio de México, 1974). In *A Spanish-Mexican Peasant Community: Arandas in Jalisco, Mexico* (Berkeley: University of California Press, 1933), 36–40, Paul Schuster Taylor discusses the experiences of rural villagers who had left the Mexican central plateau for the United States during the Cristero Rebellion and later returned to their home village.
16. Gabriela F. Arredondo, "'What! The Mexicans, Americans?': Race and Ethnicity, Mexicans in Chicago, 1916–1939" (Ph.D. diss., University of Chicago, 1999), 38–41; Manuel Bravo"Interview by Jesse J. Escalante," in *Jesse Escalante Oral Histories*; Serfafin García"Interview by Jesse J. Escalante," in *Jesse Escalante Oral Histories*; Anne M. Martínez, "Bordering on the Sacred: Religion, Nation, and U.S.-Mexican Relations, 1910–1929" (Ph.D. diss., University of Minnesota, 2003), 43–44. In their recorded interviews, both García and Bravo briefly discussthe role of the Cristero Rebellion in a renewed influx of Mexican immigrants to South Chicago.
17. García, interview by Escalante.
18. For more on the importance of investigating macroeconomic factors while studying the motivations of individuals and/or small groups, see Tamar Diana Wilson, "Theoretical Approaches to Mexican Wage Labor Migration," *Latin American Perspectives* 20, no. 3 (Summer 1993): 119.
19. Elizabeth A. Hughes, *Living Conditions for Small-Wage Earners in Chicago* (Chicago: City of Chicago Department of Public Welfare, 1925), 10. The entire pool of heads-of-households represented sixteen Mexican states.
20. Edward Jackson Baur, "Delinquency Among Mexican Boys in South Chicago" (M.A. thesis, University of Chicago, 1938), 21.
21. Mae M. Ngai, *Impossible Subjects: Illegal Aliens and the Making of Modern America, Politics and Society in Twentieth-Century America* (Princeton: Princeton University Press, 2004), 17–20.
22. Louis Bloch, "Facts About Mexican Immigration Before and Since the Quota Restriction Laws," *Journal of the American Statistical Association* 24, no. 165 (1929), 50; Ngai, *Impossible*

Subjects, 21–25. The 1924 immigration act set a minimum quota of 110 per year for any nationality. For a more comprehensive examination of the Immigration Law of 1924 and the nativist motivations behind it , see Ngai, *Impossible Subjects*, 17–55. Ngai, on pages 52–53, discusses, albeit briefly, the anti-Mexican immigration debate and "clamor" to have Mexicans included in the quota law. For more on this post-quota law and anti-Mexican debate, see Clare Sheridan, "Contested Citizenship: National Identity and the Mexican Immigration Debates of the 1920s," *Journal of American Ethnic History* 21, no. 3 (Spring 2002): 5–8.

23. Matthew Frye Jacobson, *Whiteness of a Different Color: European Immigrants and the Alchemy of Race* (Cambridge: Harvard University Press, 1998), 8.
24. Ibid., 9.
25. Arredondo, *Mexican Chicago*, 38.
26. Ngai, *Impossible Subjects*, 52–53.

CHAPTER 2

1. Gilbert Martínez, "Interview by Jesse J. Escalante," in *Jesse Escalante Oral Histories, Global Communities Collection*, Chicago History Museum.
2. Lucio Franco"Interview by Jesse J. Escalante," in *Jesse Escalante Oral Histories*. Lucio Franco Border Card, March 17, 1923, National Archives and Records Administration (NARA), Washington, D.C., Nonstatistical Manifests and Statistical Index Cards of Aliens Arriving at Laredo, Texas, May 1903–November 1929, Record Group: 85, Records of the Immigration and Naturalization Service, Microfilm Serial A3379; Microfilm Roll 26. 1930 United States Census, Chicago, Cook, Illinois, ED 2460, p. 10B.
3. Mark Reisler, *By the Sweat of Their Brow: Mexican Immigrant Labor in the United States, 1900–1940* (Westport, CT: Greenwood, Press, 1976), 5. As early as the 1880s, small groups of U.S. farmers hired labor agents to cross into Mexico and contract Mexicans for agricultural work. See also James Slaydon, "Some Observations on Mexican Immigration," *Annals of the American Academy of Political and Social Science* 93 (1921).
4. Jose Hernandez Alvarez, "A Demographic Profile of the Mexican Immigration to the United States," *Journal of Inter-American Studies* 8, no. 3 (July 1966): 473–74; Jeffrey Marcos Garcilazo, "'Traqueros': Mexican Railroad Workers in the United States, 1870–1930" (Ph.D. diss., University of California, Santa Barbara, 1995),93–96; Zaragosa Vargas, *Proletarians of the North: A History of Mexican Industrial Workers in Detroit and the Midwest, 1917–1933* (Berkeley: University of California Press, 1993), 13–14, 19–22, 36.
5. Gunther Peck, *Reinventing Free Labor: Padrones and Immigrant Workers in the North American West, 1880–1930* (New York: Cambridge University Press, 2000), 195–97.
6. David G. Gutierrez, *Walls and Mirrors: Mexican Americans, Mexican Immigrants, and the Politics of Ethnicity* (Berkeley: University of California Press, 1995).
7. I borrow this term from David J. Weber, *Foreigners in Their Native Land; Historical Roots of the Mexican Americans* (Albuquerque: University of New Mexico Press, 1973).
8. Paul S. Taylor, *Mexican Labor in the United States: Chicago and the Calumet Region, University of California Publications in Economics* 7, no. 2 (1932): 33.
9. Robert D. Cuff, "United States Mobilization and Railroad Transportation: Lessons in Coordination and Control, 1917–1945," *Journal of Military History* 53, no. 1 (1989): 33–37.
10. Taylor, *Mexican Labor*, 32.
11. In Garcilazo, "Traqueros." Garcilazo argues that railroad companies used Mexican Americans, or "Hispanos," in New Mexico as early as 1871. Railroad company use of Mexican immigrant and Mexican-American labor in the Southwest was common by the 1880s

(74–75). For more on the use of Mexicans by railroad companies, see Vargas, *Proletarians of the North*, 34–41.

12. Cuff, "Mobilization and Railroad Transportation," 33. For more on the causes of the railroad transportation crisis, see the first three chapters of William James Cunningham, *American Railroads: Government Control and Reconstruction Policies* (Chicago: A.W. Shaw, 1922). Also, see chapters 1 and 2 of Walker D. Hines, *War History of American Railroads* (New Haven: Yale University Press, 1928).
13. Campbell Gibson and Emily Lennon, "Historical Census Statistics on the Foreign-Born Population of the United States: 1850 to 1990,"*Population Division Working Paper No. 29* (Washington, DC: U.S. Bureau of the Census, Population Division, 1999).
14. William M. Tuttle, *Race Riot: Chicago in the Red Summer of 1919*, 1st ed., Studies in American Negro Life (New York: Atheneum, 1970), 242. In addition to his extensive coverage of the 1919 Chicago race riot, Tuttle describes the racial violence in the Chicago industrial workplace and discusses the non-union reputation of black workers (108–56).
15. Gabriela F. Arredondo, "Navigating Ethno-Racial Currents: Mexicans in Chicago, 1919–1939," *Journal of Urban History* 30, no. 3 (2004): 399–400. Arredondo points out that the only reason Blanco shows up in the historical record is that he attacked a white man. González's murder is mentioned only in Mexican consular record. The U.S.-centric nature of the contemporary newspaper accounts and investigations of the riots, as well as later historical analyses by scholars, have failed to record the fact that Mexicans were involved in the violence.
16. William Z. Foster, *The Great Steel Strike and Its Lessons* (New York: B. W. Huebsch, 1920), 208.
17. For more on the relationship between African Americans and organized labor during this period, see James R Barrett, *Work and Community in the Jungle: Chicago's Packinghouse Workers, 1894–1922* (Urbana: University of Illinois Press, 1986), 208–24; Sterling Denhard Spero and Abram Lincoln Harris, *The Black Worker: The Negro and the Labor Movement* (New York: Columbia University Press, 1931, reprint, New York: Atheneum, 1968); Tuttle, *Race Riot*, 108–56.
18. David Brody, *Labor in Crisis: The Steel Strike of 1919*, Critical Periods of History (Philadelphia,: Lippincott, 1965), 162–63. This classic history surveys the nationwide development and effects of the 1919 steel strike. A product of an older labor history that barely acknowledged African Americans as anything more than strikebreakers, *Labor in Crisis* did not mention the entrance of Mexicans into the industry during the strike. Brody quoted a union official's report of "Chicago conditions" from the December 13–14, 1919 minutes of the National Committee for Organizing Iron and Steel Workers of the AFL; the official noted the "bad effect on morale of the white men to see blacks crowding into the mills to take their jobs" (163, 200 n.21). In fact, Foster contended that African-American leaders welcomed the institutional prejudice of labor unions as an excuse for directing the members of their communities to be strikebreakers, allowing them entry into a previously closed occupation. Foster, *The Great Steel Strike and its Lesson*, 210–11.
19. Ibid.; Lizabeth Cohen, *Making a New Deal: Industrial Workers in Chicago, 1919–1939* (New York: Cambridge University Press, 1990), 42–43; Kenneth Warren, *Big Steel: The First Century of the United States Steel Corporation* (Pittsburgh: University of Pittsburgh Press, 2001), 117.
20. Brody, *Labor in Crisis*, 163.
21. Armandaez, interview by Paul S. Taylor, 1928, box 11, file 46, *Paul Schuster Taylor Papers*, BANC MSS 84/38 c, The Bancroft Library, University of California, Berkeley. Despite

copious reportage of the 1919 steel strike in the New York Times and the Chicago Tribune, there is no significant mention of Mexicans hired as strikebreakers.

22. Foster, *The Great Steel Strike and Its Lesson*, 205–10.
23. Juan R. García, *Mexicans in the Midwest, 1900–1932* (Tucson: University of Arizona Press, 1996), 38–40; Vargas, *Proletarians of the North: A History of Mexican Industrial Workers in Detroit and the Midwest, 1917–1933*, 57, 64; Arredondo, *Mexican Chicago: Race, Identity, and Nation, 1916–39*; Cohen, *Making a New Deal*, 41, 46. Cohen lists Mexicans along with African Americans as strikebreakers without further discussion of their involvement or motivations.
24. D. P. Thompson interview by Paul S. Taylor, 1928, Box 11, File 32, *Paul Schuster Taylor Papers*.
25. Clyde M. Brading interview by Paul S. Taylor, 1928, Box 11, File 32, *Paul Schuster Taylor Papers*.
26. Ibid.
27. Ibid.
28. Neil Foley, *The White Scourge: Mexicans, Blacks, and Poor Whites in Texas Cotton Culture* (Berkeley: University of California Press, 1997), 6.
29. According to Foley, the "white scourge" were white tenant farmers who lost economic status and had, in the eyes of those in the dominant society, become "sorry whites" and less masculine than other whites. For more on the historical use of the term "white scourge," see Foley, *The White Scourge*.
30. Foley, *The White Scourge*, 38–39. In this quotation, Foley is referring to Mexicans in the Texas cotton industry. While examining poor white tenant farmers, Foley argues that "Whites worried that their race had become vulnerable to 'pollution' by the growing population of poor whites on cotton farms and not-so-white Mexicans" (39).
31. B. C. McLeod interview by Paul S. Taylor, 1928, Box 11, File 32, *Paul Schuster Taylor Papers*.
32. Richard J. Wuerst interview by Paul S. Taylor, 1928, Box 11, File 32, *Paul Schuster Taylor Papers*. Taylor identifies the manager as Mr. Wuerst. Wuerst's full name is identified via Selective Service Registration Cards, World War II: Fourth Registration National Archives and Records Administration (NARA); Washington, DC; State Headquarters: Illinois; Microfilm Series: M2097; Microfilm Roll: 321; 1930 U.S. Census, Chicago, Illinois, ED 16-1454, p. 11A.
33. Gabriela F. Arredondo, *Mexican Chicago: Race, Identity, and Nation, 1916–39* (Urbana: University of Illinois Press, 2008), 17. For more on the third space in *Mexicano* communities, see David G. Gutierrez, "Migration, Emergent Ethnicity, and the 'Third Space': The Shifting Politics of Nationalism in Greater Mexico," *Journal of American History* 86, no. 2 (1999).

CHAPTER 3

1. Serafin García, "Interview by Jesse J. Escalante," *Jesse Escalante Oral Histories, Global Communities Collection*, Chicago History Museum.
2. Justino Cordero, "Interview by Jesse J. Escalante," in *Jesse Escalante Oral Histories*.
3. 1920 U.S. Census, Bastrop, Texas, ED 16, p. 6A.
4. Tamar Diana Wilson, "Theoretical Approaches to Mexican Wage Labor Migration," *Latin American Persepctives* 20, no. 3 (Summer 1993): 109.
5. For more on the Great African-American Migration to Chicago, see James R. Grossman, *Land of Hope: Chicago, Black Southerners, and the Great Migration* (Chicago: University of Chicago Press, 1989).

6. For further discussion of Mexican chain and circular migration to the United States, see Gunther Peck, "Reinventing Free Labor: Immigrant Padrones and Contract Laborers in North America, 1885–1925," *Journal of American History* 83, no. 3 (1996): 43–44; Vicki L. Ruiz, *From Out of the Shadows: Mexican Women in Twentieth-Century America* (Oxford: Oxford University Press, 1998), 7–8; George J. Sánchez, *Becoming Mexican American: Ethnicity, Culture, and Identity in Chicano Los Angeles, 1900–1945* (New York: Oxford University Press, 1993), 41–45,133–36; Paul Schuster Taylor, *A Spanish-Mexican Peasant Community; Arandas in Jalisco, Mexico* (Berkeley: University of California Press, 1933), 35–63; Zaragosa Vargas, *Proletarians of the North: A History of Mexican Industrial Workers in Detroit and the Midwest, 1917–1933* (Berkeley: University of California Press, 1993), 21–34, 84.
7. For a discussion on post-annexation identity, see Arnoldo De León, *Ethnicity in the Sunbelt: Mexican Americans in Houston*, 1st Texas A&M University Press ed. (College Station: Texas A&M University Press, 2001); Mario T. García, *Desert Immigrants: The Mexicans of El Paso, 1880–1920* (New Haven: Yale University Press, 1981); Emilio Zamora, *The World of the Mexican Worker in Texas* (College Station: Texas A & M University Press, 1993). For more on the migration theory relating to the migration of Mexicans to the United States, Douglas S. Massey et al., "An Evaluation of International Migration Theory: The North American Case," *Population and Development Review* 20, no. 4 (December 1994); Wilson, "Mexican Wage Labor Migration." See also Ruiz, *From Out of the Shadows*, 7.
8. John Bodnar, *The Transplanted: A History of Immigrants in Urban America* (Bloomington: Indiana University Press, 1985), 58; Peck, "Reinventing Free Labor," 851.
9. Peck, "Reinventing Free Labor," 851.
10. "Betabeleros," *Mexico*, Chicago, April 4, 1925.
11. Paul S. Taylor, "Canal Street Employment District," 1928, in *Paul Schuster Taylor Papers*, BANC MSS 84/38 c, The Bancroft Library, University of California, Berkeley. Taylor interviewed several workers waiting for work in the Canal Street Employment District. The workers, going over the job flyers when interviewed, were willing to take work as track workers or picking sugar beets.
12. García, "Interview by Jesse J. Escalante." See David A. Badillo, *Latinos in Michigan* (East Lansing: Michigan State University Press, 2003), 5–6, for a discussion of the *betabelero* community in Michigan during this period.
13. Mrs. Kembell, "Interview by Paul S. Taylor,"1928, in *Paul Schuster Taylor Papers.*
14. Ibid.
15. Pierrette Hondagneu-Sotelo, *Gendered Transitions: Mexican Experiences of Immigration* (Berkeley: University of California Press, 1994), 7.
16. Luis García, "Interview by Paul S. Taylor,"1928, in *Paul Schuster Taylor Papers.*
17. U.S. Census, Chicago, IL, ED 16-2458, p. 7B.
18. Ernastine Barrato and Teresa Santos, "Interview by Jesse J. Escalante," in *Jesse Escalante Oral Histories*; 1930 U.S. Census, Chicago, IL, ED 16-2458 p. 12A.
19. Born in Seguin, Texas, Rios's father was a U.S. citizen and Texan of Mexican descent.
20. Mercedes Rios Radica, "Interview by Jesse J. Escalante," in *Jesse Escalante Oral Histories.*
21. García, "Interview by Jesse J. Escalante."
22. Ibid.
23. Jose S. Rodriguez, "Interview by Paul S. Taylor," 1928, in *Paul Schuster Taylor Papers.*
24. 1920 U.S. Census, Chicago, IL, ED 1088 p. 1A.
25. Francisco Huerta, "Interview by Paul S. Taylor," box 11, file 32, *Paul Schuster Taylor Papers.* Although Huerta did not live in South Chicago, his story is significant because of his

journey to Chicago, his boxcar experience, and the fact that he came to be respected by Mexicans citywide as a journalist and business owner.

26. Edward Jackson Baur, "Delinquency among Mexican Boys in South Chicago" (M.A. thesis, University of Chicago, 1938), 21; Elizabeth Ann Hughes, "Living Conditions for Small Wage Earners in Chicago" (Chicago: City of Chicago Department of Public Welfare, 1925), 10.
27. For more on how railroads changed the Mexican immigrants' journey, see Jeffrey Marcos Garcilazo, "'Traqueros': Mexican Railroad Workers in the United States, 1870–1930" (Ph.D. diss., University of California, Santa Barbara, 1995, 45–48, 67–107.
28. Baur, "Delinquency," 21; Hughes, "Living Conditions," 10; Robert Redfield, "The Antecedents of Mexican Immigration to the United States," *American Journal of Sociology* 35, no. 3 (1929), 435. See also Mark Reisler, *By the Sweat of Their Brow: Mexican Immigrant Labor in the United States, 1900–1940* (Westport, CT: Greenwood Press, 1976), 17.
29. John E. Bodnar, *The Transplanted: A History of Immigrants in Urban America* (Bloomington: Indiana University Press, 1985), 68.
30. For more on the United States as a safety valve and source of inspiration, especially in the Southwest, see Ruíz, *From Out of the Shadows: Mexican Women in Twentieth-Century America*, 8–9.
31. Bodnar, *The Transplanted*, 68.
32. For the purposes of this study, the American Southwest is defined as Texas, New Mexico, Arizona, and the Southern California.
33. For more on the importance of the fact that no pre-existing Mexican community existed at the time of this migration, and how that dramatically differentiates the American Midwest from the American Southwest, see Louise Año Nuevo Kerr, "'The Chicano Experience in Chicago: 1920–1970" (Ph.D. diss., University of Illinois at Chicago Circle, 1976); and Gabriela F. Arredondo, *Mexican Chicago : Race, Identity, and Nation, 1916–39* (Urbana: University of Illinois Press, 2008), 18–20.

CHAPTER 4

1. Gabriela F. Arredondo, *Mexican Chicago: Race, Identity, and Nation, 1916–39* (Urbana: University of Illinois Press, 2008), 16.
2. Nick Svalina, "Interview by Jesse J. Escalante," in *Jesse Escalante Oral Histories, Global Communities Collection*, Chicago History Museum; 1930 U.S. Census, Chicago, IL, ED 2458, p. 38A; Sam Svalina Naturalization Card, Soundex Index to Naturalization Petitions for the United States District and Circuit Courts, Northern District of Illinois and Immigration and Naturalization Service District 9, 1840–1950, Microfilm Serial M1285, Microfilm Roll 150.
3. Peter Thomas Alter, "Mexicans and Serbs in Southeast Chicago: Racial Group Formation during the Twentieth Century," *Journal of the Illinois State Historical Society* 94, no. 4 (2001): 403–404.
4. David R. Roediger, *The Wages of Whiteness: Race and the Making of the American Working Class* (New York: Verso, 1999), 141–42. As early as the 1840s, Mexicans were not only given a "less-than-white" status in popular American political culture, but were used as justification to elevate most European immigrants to "white" status. George Lipsitz concurs that the creation of racial hierarchies with "whiteness" at the pinnacle happened through deliberate strategies and not by happenstance. He further argues that these efforts to imbue European Americans with an emotional investment in being white have occurred

from the colonial era to the present. George Lipsitz, "The Possessive Investment in Whiteness: Racialized Social Democracy and the "White" Problem in American Studies," *American Quarterly* 47, no. 3 (1995): 371–80. For more on the ethnic and racial classification of immigrants, see James R. Barrett and David Roediger, "Inbetween Peoples: Race Nationality and the `New Immigrant' Working Class," *Journal of American Ethnic History* 16, no. 3 (1997). For a historical discussion of "whiteness" in the United States, see especially Matthew Frye Jacobson, *Whiteness of a Different Color: European Immigrants and the Alchemy of Race* (Cambridge: Harvard University Press, 1998), 39–135. See also David R. Roediger, *Working toward Whiteness: How America's Immigrants Became White: The Strange Journey from Ellis Island to the Suburbs* (New York: Basic Books, 2005).

5. Isabel García, Fernando Guiuliani, and Esther Wiesenfeld, "Community and Sense of Community: The Case of an Urban Barrio in Caracas," *Journal of Community Psychology* 27, no. 6 (1999): 729.
6. Dominic A. Pacyga and Ellen Skerrett, *Chicago, City of Neighborhoods: Histories and Tours* (Chicago: Loyola University Press, 1986), 409–10; Louis Wirth and Eleanor H. Bernert, *Local Community Fact Book of Chicago* (Chicago: University of Chicago Press, 1932), 46.
7. Brosch, David, "The Historical Development of Three Chicago Millgates: The East Side," in *The Historical Development of Three Chicago Millgates: South Chicago, East Side, South Deering*, eds. Marcia Kijewski, David Brosch, and Robert Bulanda (Chicago: Illinois Labor History Society, ca. 1973), 4.
8. Jorge Hernandez-Fujigaki, "Mexican Steelworkers and the United Steelworkers of America in the Midwest: The Inland Steel Experience, 1936–1976," (Ph.D. diss., University of Chicago, 1991), 27. The specific ethnic groups hired at Illinois Steel switched from English, German, Welsh, Belgian, and Dutch to Polish, Bohemian, Slovak, Serbian, Bulgarian, Lithuanian, Magyar, Romanian, and Macedonian.
9. Hernandez-Fujigaki, "Mexican Steelworkers and the USWA," 27–32; Paul S. Taylor, *Mexican Labor in the United States: Chicago and the Calumet Region, University of California Publications in Economics* 7, no. 2 (1932): 42–44. See pages 27–34 of Hernandez-Fujigaki for a more detailed examination of the early racial and ethnic make-up of the early South Chicago, East Chicago, and Gary steel industry workforce.
10. James R. Grossman, *Land of Hope: Chicago, Black Southerners, and the Great Migration* (Chicago: University of Chicago Press, 1989), 113.
11. John B. Appleton, "The Iron and Steel Industry in the Calumet District: A Study in Economic Geography," *University of Illinois Studies in the Social Sciences* 8, no. 2 (1925): 12.
12. Sophenisba P. Breckinridge and Edith Abbott, "Chicago Housing Conditions: V: South Chicago at the Gates of the Steel Mills," *American Journal of Sociology* 17, no. 2 (1911): 174. On the development of a "distinctive urban sensibility toward nature" in Chicago during the end of the nineteenth century, see Jonathan J. Keyes, "Urbs in Horto: Chicago and Nature, 1833–1874" (Ph.D. diss., University of Chicago, 2003). Keyes argues that the stockyards and other industrial areas were at one end of the range of perspectives on the place of nature in the city, while the park system stood at the other. However, both were equally attempts to solve the problems of the industrial urban environment.
13. Breckinridge and Abbott, "At the Gates," 174.
14. Ibid.
15. Sylvia Hood Washington, *Packing Them In: An Archaeology of Environmental Racism in Chicago, 1865–1954* (Lanham, MD: Lexington Books, 2005), 18–19. Washington argues that

"environmental racism" as a term "embraces environmental disenfranchisement based upon multiple race categories (including whites), class, and ethnicity," 21.

16. Ibid., 78.
17. Mary Faith Adams, "Present Housing Conditions in South Chicago, South Deering and Pullman" (M.A. thesis, University of Chicago, 1926), 62–63; Elizabeth A. Hughes, *Living Conditions for Small-Wage Earners in Chicago* (Chicago: City of Chicago Department of Public Welfare, 1925), 8–9; David S. Weber, "Anglo Views of Mexican Immigrants: Popular Perceptions and Neighborhood Realities in Chicago, 1900–1940" (Ph.D. diss., Ohio State University, 1982), 114. Weber describes some of the physical "zones of transition" Mexicans confronted in Chicago.
18. Hughes, "Living Conditions," 9.
19. Robert Redfield, "Robert Redfield Journal," box 59, folder 2, *Robert Redfield Papers*, Special Collections Research Center, University of Chicago Library; Dorothea Kahn, "Mexicans Bring Romance to Drab Part of Chicago in Their Box-Car Villages: 30,000 Now Make City," *Christian Science Monitor*, May 23, 1931.
20. Kahn, "Mexicans Bring Romance"
21. Jeffrey Marcos Garcilazo provides a thorough description of various Mexican boxcar camps throughout the country in "'Traqueros': Mexican Railroad Workers in the United States, 1870–1930" (Ph.D. diss., University of California, Santa Barbara, 1995), 229–76. Garcilazo argues that generalizations about boxcar colonies are almost impossible to make given the "diversity of experiences around the country" (229). Using interviews and descriptions from the Paul S. Taylor Collection, Garcilazo also gives examples of Chicago-area boxcar camps (255–59).
22. Paul Frederick Cressey, "The Succession of Cultural Groups in the City of Chicago" (Ph.D. diss., University of Chicago, 1930), 152.
23. South Chicago Resident, interview by Paul S. Taylor, box 11, file 32, *Paul Schuster Taylor Papers*, BANC MSS 84/38 c, The Bancroft Library, University of California, Berkeley.
24. Ibid.
25. Ibid.
26. Justino Cordero, "Interview by Jesse J. Escalante," in *Jesse Escalante Oral Histories.*
27. 1920 U. S. Census, Chicago, IL, ED 503, p. 9B.
28. 1920 U. S. Census, Chicago, IL, ED 504, p. 8B.
29. 1920 U. S. Census, Chicago, IL, ED 500, p. 5B.
30. 1920 U. S. Census, Chicago, IL, ED 504, p. 8B.
31. 1920 U. S. Census, Chicago, IL, ED 509, p. 4B.
32. 1920 U. S. Census, Chicago, IL, ED 509, p.12A.
33. 1920 U. S. Census, Chicago, IL, ED 501, p. 2A.
34. Alfredo de Avila, "Interview by Jesse J. Escalante," in *Jesse Escalante Oral Histories.*
35. Ibid.; Alfredo De Avila Entry Card, Nonstatistical Manifests and Statistical Index Cards of Aliens Arriving at El Paso, Texas, 1905–1927, Record Group 85, Records of the Immigration and Naturalization Service, Microfilm Serial A3406, Microfilm Roll 9.
36. Hughes, "Living Conditions," 12–13.
37. 1920 U. S. Census, Chicago, IL, ED 503, pp. 4A, 4B, 8B, 9A, 9B, 10A, 11B; 1930 U. S. Census, Chicago, IL, ED 2459, pp. 4B, 5A, 7A, 7B, 8A, 8B; 1940 U. S. Census, Chicago, IL, ED 663, pp. 6B, 7A, 8A, 8B, 9A, 9B, 62A, 63A, 64A.
38. 1920 U. S. Census, Chicago, IL, ED 500, pp. 3A, 4A, 4B, 5A, 5B; 1920 U. S. Census, Chicago, IL, ED 501, pp. 2B, 3A, 3B, 4A, 4B; 1930 U. S. Census, Chicago, IL, ED 2458, pp. 11B, 12A, 12B, 13A, 13B, 14B, 15A, 15B, 17B 34A, 34B, 35A, 39B, 40A, 40B, 41A, 41B; 1940 U. S.

Census, Chicago, IL, ED 661, pp. 1A, 1B, 2A, 9B, 10A, 10B; 1940 U. S. Census, Chicago, IL, ED 662, pp. 1A, 1B, 2A, 2B, 3A, 4B, 5A, 5B, 6A, 6B.

39. 1920 U. S. Census, Chicago, IL, ED 501, p. 4A.
40. 1930 U. S. Census, Chicago, IL, ED 2458, p. 1B.
41. 1940 U. S. Census, Chicago, Il, ED 662, p. 1B.
42. 1920 U. S. Census, Chicago, IL, ED 501, p. 4B.
43. 1920 U. S. Census, Chicago, IL, ED 501,p. 6B.
44. U. S. Census, Chicago, IL, ED 500, p. 5A.
45. Hughes, "Living Conditions," 7.
46. Weber, "Anglo Views," 141. Although the presence of African Americans in South Chicago kept Mexicans from the bottom of the "American" social ladder, it was South Chicago's history of bringing in and acculturating European ethnic groups that facilitated the higher degree of acceptance of Mexicans. Also, there was a lower level of institutional discrimination of Mexicans in Chicago than in the American Southwest.
47. Marcel van der Linden, "Introduction," in *Rebellious Families: Household Strategies and Collective Action in the Nineteenth and Twentieth Centuries*, ed. Jan Kok (New York: Berghahn Books, 2002), 8.
48. The significance of women's work to the household economy has been well established. For studies focusing on Mexican women, see Vicki L. Ruiz, *Cannery Women, Cannery Lives: Mexican Women, Unionization and the California Food Processing Industry, 1930–1950* (University of New Mexico Press, 1987); Patricia Zavella, *Women's Work and Chicano Families: Cannery Workers of the Santa Clara Valley* (Ithaca: Cornell University Press, 1987). On Chicago, see Arredondo, *Mexican Chicago: Race, Identity, and Nation, 1916–39*.
49. For more on the theory of household networks and personal communities, as well as specific examples of household survival strategies, see Jan Kok, *Rebellious Families: Household Strategies and Collective Action in the Nineteenth and Twentieth Centuries* (New York: Berghahn Books, 2002).
50. Mr. Fernández, interview by Paul S. Taylor, Transcript, box 11, file 32, *Paul Schuster Taylor Papers*.
51. Mr. de Gerald, interview by Paul S. Taylor, 1928, box 11, file 32, *Paul Schuster Taylor Papers*.
52. Ibid.
53. Rios Radica, "Interview by Jesse J. Escalante," *Jesse Escalante Oral Histories*.
54. Union State Bank advertisement, *Mexico*, Chicago, June 5, 1926; South Chicago Savings Bank advertisement, *El Heraldo Juvenil*, January 30, 1930, in *Mexicans in Chicago Archival Collection*, Chicago Theological Seminary.
55. Mr. de Gerald, interview by Taylor.
56. Mr. Fernández, interview by Taylor.

CHAPTER 5

1. Serafín García, "Interview by Jesse J. Escalante," in *Jesse Escalante Oral Histories, Global Communities Collection*, Chicago History Museum. Alfredo De Avila, "Interview by Jesse J. Escalante," in Jesse Escalante Oral Histories, Global Communities Collection, Chicago History Museum.
2. U.S. Census, Chicago, IL, ED 2458, p. 37A; Benigno Castillo Border Crossing Card, Nonstatistical Manifests and Statistical Index Cards of Aliens Arriving at Laredo, Texas, May 1903–November 1929, Record Group 85, Records of the Immigration and Naturalization Service; Microfilm Serial: A3379, Microfilm Roll 12.

3. Benigno Castillo, "Interview by Jesse J. Escalante," in *Jesse Escalante Oral Histories*. This interview is in Spanish. Translation by the author.
4. Ibid. For a discussion of discrimination in the workplace and the establishment of occupational patterns for Mexicans in the South Chicago and Northwest Indiana steel mills, see Jorge Hernandez-Fujigaki, "Mexican Steelworkers and the United Steelworkers of America in the Midwest: The Inland Steel Experience,1936–1976" (Ph.D. diss., University of Chicago, 1991), 53–87. Hernandez-Fujigaki argues that the "incorporation of Mexicans in the labor markets of the Southwest, primarily as menial laborers, strongly influenced the range of occupations available to them in the steel mills of the Great Lakes" (54). He also argues that "the initial occupations that Mexicans occupied in the Midwest were shaped both by their late arrival and by the perceptions of the representatives of growers, mine owners and, particularly, the railroads in the Southwest. Historically it was in the Southwest where the association between Mexicans and their lowly occupational status had been initially made" (57–58).
5. Only one Mexican was listed as a steel mill foreman in the 1920 census enumeration districts within the South Chicago neighborhood. Gerardo Rios, living as a roomer on the 9200 block of Burley Avenue, was listed as a foreman at an unnamed steel mill. He immigrated in 1914, knew English, was twenty-nine years old, and was listed as single. 1920 U.S. Census, Chicago, IL, ED 504, p. 11A.
6. Paul S. Taylor, *Mexican Labor in the United States: Chicago and the Calumet Region, University of California Publications in Economics*, vol. 7, no. 2 (Berkeley: University of California Press, 1932), 32–33.
7. Zaragosa Vargas, *Proletarians of the North: A History of Mexican Industrial Workers in Detroit and the Midwest, 1917–1933* (Berkeley: University of California Press, 1993), 49.
8. Hernandez-Fujigaki, "Mexican Steelworkers"; Vargas, *Proletarians of the North*, 48–49. South Chicago was home to a diverse group of Mexicans, Anglo Americans, African Americans, and many different ethnic Europeans. More study is necessary on the complicated racial and ethnic tensions in South Chicago. For an introduction to the issues in the city of Chicago as a whole, see Arredondo, *Mexican Chicago*.
9. Vargas, *Proletarians of the North*, 48.
10. Taylor, *Mexican Labor*, 36–37.
11. Hernandez-Fujigaki, "Mexican Steelworkers," 29–32; Taylor, *Mexican Labor*, 46.
12. Martínez, "Interview by Jesse J. Escalante."
13. Ibid.
14. Gabriela F. Arredondo, "Navigating Ethno-Racial Currents," *Journal of Urban History* 30, no. 3 (2004): 408.
15. South Chicago Resident, interview by Paul S. Taylor, box 11, file 32, *Paul Schuster Taylor Papers*.
16. Francisco Huerta, "Interview by Paul S. Taylor," box 11, file 32, *Paul Schuster Taylor Papers*.
17. Castillo, "Interview by Jesse J. Escalante." Castillo remembered that, when starting at Illinois Steel as a laborer working with a shovel "cleaning and gathering dirt," he and fellow Mexican workers made less money the first day because the other ethnic groups, primarily Polish, Irish, Croatian, and Hungarian, "didn't tell us how to start."
18. Sophonisba P. Breckinridge and Edith Abbott, "Housing Conditions in Chicago, Ill: Back of the Yards," *American Journal of Sociology* 16, no. 4 (January 1911).

19. Thomas J. Jablonsky, *Pride in the Jungle: Community and Everyday Life in Back of the Yards Chicago*, Creating the North American Landscape (Baltimore: Johns Hopkins University Press, 1993), 1–4, 33.
20. City of Chicago, *Local Community Fact Book of Chicago* (Chicago: City of Chicago, 1982), 158.
21. University of Chicago Settlement House, "Report on the Foreign Born," 1929, in *University of Chicago Settlement House Papers,* Chicago History Museum, 4.
22. Paul Frederick Cressey, "The Succession of Cultural Groups in the City of Chicago" (Ph. D. diss., University of Chicago, 1930), 152.
23. Sophonisba P. Breckinridge and Edith Abbott, "Housing Conditions in Chicago, IV: The West Side Revisited," *American Journal of Sociology* 17, no. 1 (July 1911).
24. Dominic A. Pacyga and Ellen Skerrett, *Chicago, City of Neighborhoods: Histories and Tours* (Chicago: Loyola University Press, 1986), 199.
25. Ibid.
26. University of Chicago Settlement House, "Report on the Foreign Born," 4.
27. Ibid.
28. For more on Mexican use of parks and other recreational facilities, see chapter 8 of this book.
29. Julián Samora and Richard A. Lamanna, "Mexican-Americans in a Midwest Metropolis: A Study of East Chicago," in *Forging a Community: The Latino Experience in Northwest Indiana, 1919–1975*, eds. James B. Lane and Edward J. Escobar (Chicago: Cattails Press, 1987), 138. Since they were on main railroad lines, South Chicago, East Chicago and Gary all had boxcar camps. The cities' steel mills benefited from *traqueros* who jumped their contracts. Francisco Arturo Rosales and Daniel T. Simon, "Mexican Immigrant Experience in the Urban Midwest: East Chicago, Indiana, 1919–1945," in *Forging a Community: the Latino Experience in Northwest Indiana, 1919–1975*, eds. James B. Lane and Edward J. Escobar (Chicago: Cattails Press, 1987), 137–41; Taylor, *Mexican Labor*, 36.
30. For more on Gary, Indiana, see Neil Betten and Raymond A. Mohl, "From Discrimination to Repatriation: Mexican Life in Gary, Indiana, During the Great Depression," *Pacific Historical Review* 42, no. 2 (August 1973); Powell A Moore, *The Calumet Region* (Indianapolis: Indiana Historical Bureau, 1959).
31. John B Appleton, "The Iron and Steel Industry in the Calumet District: A Study in Economic Geography," *University of Illinois Studies in the Social Sciences* 8, no. 2 (1925): 36.
32. Louise Año Nuevo Kerr, "The Chicano Experience in Chicago: 1920–1970" (Ph.D. diss., University of Illinois at Chicago Circle, 1976), 28; Ernest W. Burgess and Charles Newcomb, eds., *Census Data of the City of Chicago, 1930* (Chicago: University of Chicago Press, 1933), 28. James R. Grossman, Ann Durkin Keating, and Janice L. Reiff, *The Encyclopedia of Chicago* (Chicago: University of Chicago Press, 2004), 1043.
33. Francisco Arturo Rosales, "Mexican Immigration to the Urban Midwest during the 1920s" (Ph.D. diss., Indiana University, 1978), 171; Raymond A. Mohl and Neil Betten, "Discrimination and Repatriation," 162; Ciro Sepúlveda, "La Colonia Del Harbor: A History of Mexicanos in East Chicago, Indiana, 1919–1932" (Ph.D. diss., Notre Dame University, 1976), 3; Taylor, *Mexican Labor*, 52–53. The Mexican neighborhood of East Chicago, known as *La Colonia del Harbor*, was in the eastern section of the city of East Chicago, popularly referred to as Indiana Harbor.
34. Mike Amezcua, "The Second City Anew: Mexicans, Urban Culture, and Migration in the Transformation of Chicago, 1940–1965" (Ph.D. diss., Yale University, 2011), 110.

35. Malachy R. McCarthy, "Which Christ Came to Chicago: Catholic and Protestant Programs to Evangelize, Socialize and Americanze the Mexican Immigrant, 1900–1940" (Ph.D. diss., Loyola University of Chicago, 2002), 130.
36. "Mexicans Adopt Lincoln's Spirit," *Chicago Daily News*, August 8, 1928.
37. Immigrant Protective League, "Mexicans," 1929, in *Immigrant's Protective League Papers*, University of Illinois–Chicago Special Collections, Chicago.
38. For more about the concerns by Midwestern Mexican men about Mexican women becoming more "liberated," see Vargas, *Proletarians of the North*, 165–66.
39. Adena Miller Rich, "Educational Requirements for Naturalization—Do They Need Revision?" *Social Service Review* 17 (1944).
40. Linda Kerber first named this concept in relation to white women's roles in the early Republic. More recent scholarship has expanded the idea of republican motherhood to one of republican womanhood to include other women's roles. For more on this, see Catherine Allgor, *Parlor Politics: In Which the Ladies of Washington Help Build a City and a Government* (Charlottesville: University Press of Virginia, 2000).
41. George J. Sánchez, "'Go After the Women': Americanization and the Mexican Immigrant Woman, 1915–1929," in *Unequal Sisters: A Multicultural Reader in U.S. Women's History*, eds. Vicki L. Ruiz and Ellen Carol Dubois (New York: Routledge, 1994), 284.
42. Arredondo in *Mexican Chicago* discusses varying perceptions of the dangers of Americanization among members of the Mexican community of Chicago.
43. Paul S. Taylor, "Field Notes," in *Paul Schuster Taylor Papers*.
44. Francisco Huerta, "Interview by Paul S. Taylor," box 11, file 32, *Paul Schuster Taylor Papers*; Mercedes Rios Radica, "Interview by Jesse J. Escalante," in *Jesse Escalante Oral Histories*.
45. Mary E. Odem, *Delinquent Daughters: Protecting and Policing Adolescent Female Sexuality in the United States, 1885–1920* (Chapel Hill: University of North Carolina Press, 1995), 9.
46. Carlos Perez Lopez, "Interview by Manuel Gamio," Transcript, box 2, *Manuel Gamio Papers*, The Bancroft Library, University of California, Berkeley; Huerta, "Interview by Taylor." Perez Lopez believed that women in Chicago were more aware of their rights and were more assertive towards their husbands. He also added that "the women of better social standing are more aware of their social responsibilities and do more social work than in Mexico."
47. Huerta, "Interview by Taylor."
48. For the experiences of those ethnic European women immigrants in Chicago, see Joanne J. Meyerowitz, *Women Adrift: Independent Wage Earners in Chicago, 1880–1930* (Chicago: University of Chicago Press, 1988). On Mexican perceptions of the dangers of the dominant U.S. culture, see Vicki L. Ruiz, *From Out of the Shadows: Mexican Women in Twentieth-Century America* (Oxford: Oxford University Press, 1998), ch. 2. Arredondo, *Mexican Chicago*, ch. 4, explores these dynamics for Mexicans in Chicago as a whole.
49. Huerta, "Interview by Taylor"; Vargas, *Proletarians of the North*, 133–34. According to newspaperman Francisco Huerta, most Mexican women working in businesses outside of the home worked at the Cracker Jack factory, National Biscuit (Nabisco), the Marshall Fields mattress factory, and in needle trade shops.
50. Odem, *Delinquent Daughters*, 1.
51. Ibid.
52. Directories of general social services and recreational facilities list several Protestant-run programs and sites; Catholic recreation programs specifically for Mexicans did not appear in South Chicago until the late 1930s. Año Nuevo Kerr, "Chicano Experience," 57–58;

Henry Seymour Brown et al., "Report of the Committee on Findings set up by the Comity Commission of the Chicago Federation of Churches on the Survey of South Chicago by the Department of Research and Survey under the Direction of Professor Holt" (Chicago: Chicago Historical Society, 1928); City of Chicago, *Leisure Time Directory* (Chicago: 1940); Arthur J. Todd, William F. Byron, and Howard L. Vierow, *The Chicago Recreation Survey: Private Recreation*, 5 vols., vol. 3 (Chicago: Chicago Recreation Commission and Northwestern University, 1937). Contemporary social work master's theses such as Anita Edgar Jones, "Conditions Surrounding Mexicans in Chicago" (M.A. thesis, University of Chicago, 1928), and Marian Lorena Osborn, "The Development of Recreation in the South Park System of Chicago" (M.A. thesis, University of Chicago, 1928), also document the prevalence of Protestant neighborhood recreation facilities. The Protestant churches in Chicago commissioned neighborhood surveys in 1933; for South Chicago see Chicago Council of Social Agencies, *Social Service Directory and Yearbook* (Chicago: Chicago Council of Social Agencies, 1933).

53. Robert A. Orsi, *Thank You, St. Jude: Women's Devotion to the Patron Saint of Hopeless Causes* (New Haven, CT: Yale University Press, 1996), 3.
54. Ibid., 4; McCarthy, "Which Christ Came to Chicago," 129.
55. McCarthy, "Which Christ Came to Chicago," 152.
56. Ibid., 204–206; Orsi, *Thank You, St. Jude*, 4.
57. McCarthy, "Which Christ Came to Chicago," 8.
58. Arredondo, *Mexican Chicago: Race, Identity, and Nation, 1916–39*; Vicki Ruiz, "'Star Struck': Acculturation, Adolescence, and the Mexican American Woman, 1920–1950," in *Building with Our Hands: New Directions in Chicana Studies*, eds. Adela de la Torre and Beatriz Pesquera (Berkeley: University of California Press, 1993).
59. Socorro Zaragoza, "Interview by Paul S. Taylor," box 11, file 32, *Paul Schuster Taylor Papers.*
60. Ibid.
61. Ignacio Vallarta, "Interview by Paul S. Taylor," box 11, file 32, *Paul Schuster Taylor Papers.*
62. Lacy Simms, "Interview by Paul S Taylor," box 11, file 32, *Paul Schuster Taylor Papers.*
63. Huerta, "Interview by Taylor."
64. Peggy Pascoe, *What Comes Naturally: Miscegenation Law and the Making of Race in America* (New York: Oxford University Press, 2009), 2.
65. Ibid., 313.
66. Ibid.
67. Manuel Gamio, "Linguistic Phenomenon: Accepted Anglicisms in Everyday Spanish in Chicago," 1926, *Manuel Gamio Papers.*
68. McCarthy, "Which Christ Came to Chicago," 6–7.
69. "We Must Give our Children the Love of our *Patria*," *Mexico*, Chicago, April 11, 1925.
70. Ibid. Since Americanization means different things to different organizations, a "one-size-fits-all" definition is hard to come by. For the purposes of this book, I borrow Malachy McCarthy's definition that is based on the work of Jim Barrett and Paula Kane: "Americanization is a process undertaken by individuals or immigrant groups, whose goal is to encourage immigrant adjustment to the United States, whether that be the gradual acceptance of its democratic ideology, cultural norms, social behavior or Protestant religious belief (McCarthy, "Which Christ Came to Chicago," 7).
71. *Correo Mexicano*, "And Now Men," 2.
72. Taylor, *Mexican Labor*, 175.

73. For a discussion on the debate around learning English in Los Angeles, see chapter 4 of Sánchez, *Becoming Mexican American: Ethnicity, Culture, and Identity in Chicano Los Angeles, 1900–1945*.
74. Juan R. García, *Mexicans in the Midwest, 1900–1932* (Tucson: University of Arizona Press, 1996), 216; Sánchez, *Becoming Mexican American*, 101.
75. Studies of language literacy among current immigrant populations have argued that refusing to learn English can be either an unconscious or conscious act of resistance against demands by the dominant culture. See, for example, Adrian Blackledge, "Language, Literacy, and Social Justice: The Experiences of Bangladeshi Women in Birmingham, UK," *Journal of Multilingual and Multicultural Development* 20, no. 3 (1999). Although this article focuses on a different population, the general point is applicable to other immigrant groups.
76. "We Must Give our Children the Love of our *Patria*," *Mexico.*
77. Augustin J. Fink, "Interview by Manuel Gamio," June 26, 1926, Box 3, File 11, *Manuel Gamio Papers*; Taylor, *Mexican Labor*, 149–55. 1930 U.S. Census, Chicago, IL, ED 1711, p. 12B; Augustin Fink Border Crossing Card, Manifests of Statistical and Some Nonstatistical Alien Arrivals at Laredo, Texas, May 1903–April 1955, Record Group: 85, Records of the Immigration and Naturalization Service, Microfilm Serial A3437, Microfilm Roll 77.
78. Juan B. Medina, "Interview by Manuel Gamio," June 21, 1926, Transcript, Box 3, File 11, *Manuel Gamio Papers*; 1930 U.S. Census, Chicago, IL, ED 1642, p. 10A; Juan B Medina World War I Draft Card, World War I Selective Service System Draft Registration Cards, 1917–1918, M1509, Registration Location: Cook County, Illinois, Roll 1613515, Draft Board 40.
79. Robert C. Jones, "Research Notes of Robert Jones," in *Mexicans in Chicago Archival Collection,* Chicago Theological Seminary, ca. 1930.
80. García, *Mexicans in the Midwest, 1900–1932*, 155.

CHAPTER 6

1. Jose Vasconcelos, "Speech before the mutualista Ignacio Zaragoza at Hull House, Chicago," June 2, 1928, *Paul Schuster Taylor Papers*, BANC MSS 84/38 c, The Bancroft Library, University of California, Berkeley. This quotation is a translation by Paul S. Taylor.
2. Ibid. "Speech at Hull House."
3. Jesse Parez, "Interview by unknown researcher," 1945, *Foreign Populations Collection*, Chicago History Museum, Chicago.
4. Luis Felipe Recinos, "Life Histories: The Arce Family in the United States," 1927, *Manuel Gamio Papers*, The Bancroft Library, University of California, Berkeley
5. Max Guzman, "Interview by Jesse J. Escalante," in *Jesse Escalante Oral Histories, Global Communities Collection,* Chicago History Museum; Recinos, "The Arce Family"; Paul S. Taylor, *Mexican Labor in the United States: Chicago and the Calumet Region, University of California Publications in Economics,* vol. 7, no. 2 (1932), 96.
6. Robert Redfield, "Typed Field Notes of United Charities Cases," 1929, Box 59, Folder 2, in *Robert Redfield Papers,* University of Chicago Special Collections, Chicago.
7. For more on the concept of continental citizenship, see Gabriela Arredondo, *Mexican Chicago: Race, Identity, and Nation, 1916–39* (Urbana: University of Illinois Press, 2008), 172.
8. Recinos, "The Arce Family."
9. Arredondo, *Mexican Chicago*, 7.

10. Lacy Simms, "Interview by Paul S. Taylor," June 1928, *Paul Schuster Taylor Papers*, BANC MSS 84/38 c, The Bancroft Library, University of California, Berkeley; 1920 U.S. Census, Jicarilla, New Mexico, ED 101, p. 11B; 1930 U.S. Census, Highland Park, Illinois, ED 27, p. 16A. Contemporary scholars offered similar reasons. See especially Edward Jackson Baur, "Delinquency Among Mexican Boys in South Chicago" (M.A. thesis, University of Chicago, 1938); Norman D. Humphrey, "The Detroit Mexican Immigrant and Naturalization," *Social Forces* 22, no. 3 (1944); and Helen W. Walker, "Mexican Immigrants and American Citizenship," *Sociology and Social Research* 13 (1929).
11. F. Arturo Rosales, *Pobre Raza!: Violence, Justice and Mobilization among Mexico Lindo Immigrants, 1900–1936* (Austin: University of Texas Press, 1999), 195.
12. John A García, "Political Integration of Mexican Immigrants: Explorations into the Naturalization Process," *International Migration Review* 15, no. 4 (1981): 612; Leo Grebler, "The Naturalization of Mexican Immigrants in the United States," *International Migration Review* 1, no. 1 (1966): 22; George J. Sánchez, *Becoming Mexican American: Ethnicity, Culture and Identity in Chicano Los Angeles, 1900–1945* (New York: Oxford University Press, 1993), 105.
13. Taylor, *Mexican Labor*, 214.
14. Robert C. Jones and Louis R. Wilson, *The Mexican in Chicago, The Racial and Nationality Groups of Chicago: Their Religious Faiths and Conditions* (Chicago: Comity Commission of the Chicago Church Federation, 1931), 27.
15. Ibid.
16. Vasconcelos, *Speech at Hull House*.
17. Jones and Wilson, *The Mexican in Chicago*, 28.
18. Ibid, 27.
19. Robert C. Jones, "Research Notes of Robert Jones," in *Mexicans in Chicago Archival Collection*, Chicago Theological Seminary, 13–14.
20. Ibid., 117.
21. Unnamed Polish Boys, "Interview by Paul S. Taylor," 1928, *Paul Schuster Taylor Papers*.
22. 1930 U.S. Census, Chicago, IL, ED 216, pp. 20B, 21B, 24A, 24B; 1930 U.S. Census, Chicago, IL, ED 319, p. 34A.
23. "A Doctor Makes Insolent Declarations," *Mexico*, October 20, 1928, in *Foreign Language Press Survey* (*FLPS*).
24. Ibid.
25. Ibid.
26. Eugenics is the study of methods of improving the quality of the human race, usually by "selective breeding." This primarily manifested itself in the United States through state-sponsored forced sterilization of people whom the state classified as "unfit." Others, like Dr. Goldberg, wanted to minimize contact between what he judged to be inferior peoples and white, non-immigrant Americans. For more on the eugenics movement in the United States, see Edwin Black, *War against the Weak: Eugenics and America's Campaign to Create a Master Race*, expanded edition (New York: Dialog Press, 2012).
27. "A Doctor Makes Insolent Declarations."
28. The *Tribune* article that is the basis of the *Mexico* article is "Depicts Mexican Immigrants as Health Menace," *Chicago Tribune*, October 19, 1928.
29. Godias J. Drolet, "Discussion of Tuberculosis in Mexicans," *American Journal of Public Health and the Nation's Health* 9, no. 3 (March 1929); Benjamin J. Goldberg, "Tuberculosis in Racial Types with Special Reference to Mexicans," *American Journal of Public Health and*

the Nation's Health 9, no. 3 (March 1929). Goldberg's paper provides a fascinating snapshot of eugenics. Within the paper, Goldberg discusses European superiority and the "civilized" status of its people as a result of their having gone through "survival of the fittest" through the various plagues suffered by Europeans. He is careful to make the Irish and "maybe" the Swedes an exception to this European civilized status because they "have not, perhaps, passed all the way along the trail." This plays right into the idea that the Irish are just starting to become white but not yet there. He also labels the "negro" as having much of the same problems as Mexicans because of their isolation in the South, but acknowledges the responsibility to protect African Americans because they are here to stay. He then gives a history of tuberculosis in "The Primitive Races: Indians, Negros, and Mexicans."

30. Paul S. Taylor, "Field Notes," 1928, *Paul Schuster Taylor Papers.*
31. Taylor, *Mexican Labor*, 96.
32. Ibid., 235. Taylor provides excerpts of the interview with the settlement house worker.
33. James C. Scott, "Resistance without Protest and without Organization: Peasant Opposition to the Islamic Zakat and the Christian Tithe," *Comparative Studies in Society and History* 29, no. 3 (1987); James C. Scott, *Weapons of the Weak: Everyday Forms of Peasant Resistance* (New Haven: Yale University Press, 1985); James C. Scott and Benedict J. Tria Kerkvliet, "Everyday Forms of Peasant Resistance in South-East Asia," *Journal of Peasant Studies* 13, no. 2 (1986); quotation from Scott, "Resistance without Protest and without Organization," 419–20. For the argument that Scott swung too far towards individual actions, see for example Matthew C. Guttmann, "Rituals of Resistance: A Critique of the Theory of Everyday Forms of Resistance," *Latin American Perspectives* 20, no. 2 (1993). An earlier study on workers' resistance and their ability to control productivity is Stanley Bernard Mathewson, *Restriction of Output among Unorganized Workers* (New York: Viking Press, 1931).
34. Sherry B Ortner, "Resistance and the Problem of Ethnographic Refusal," *Comparative Studies in Society and History* 37, no. 1 (1995), provides a survey of these critiques in the context of defining resistance studies.
35. Anthropologists in particular have been quite taken with Scott's work, whether as a foundation or a goad to critique. K. Sivaramakrishnan, "Introduction to 'Moral Economies, State Spaces, and Categorical Violence,'" *American Anthropologist* 107, no. 3 (2005), surveys Scott's monographs, while Sivaramakrishnan's article "Some Intellectual Genealogies for the Concept of Everyday Resistance," in the same issue, considers the influence Scott's work has had on anthropological literature on resistance.
36. These works range across time periods and groups. Historians of race and ethnicity in the United States have more often turned to Robin D. G. Kelley, *Race Rebels: Culture, Politics, and the Black Working Class* (New York: Free Press, 1996), which is a more recent formulation of Scott's theory about unarticulated resistance, rather than directly to Scott's work. Remembering acts of resistance in his own youth, Kelley recounted that while working at a McDonald's, he and his fellow employees were "constantly inventing new ways to rebel, ways rooted in our own peculiar circumstances." Not only did Kelly and his friends not know "where the struggle would end," but he didn't consider himself part of a movement; "indeed, I doubt any of us thought we were part of a movement" (3). Monographs and essays on race and ethnicity in the twentieth-century that use Kelley include Carl Husemoller Nightingale, "The Global Inner City: Toward a Historical Analysis," in *W.E.B. Du Bois, Race, and the City: The Philadelphia Negro and Its Legacy*, eds. Michael B. Katz and Thomas J. Sugrue (Philadelphia: University of Pennsylvania Press, 1998); Kimberley

L. Phillips, *Alabama North: African-American Migrants, Community, and Working-Class Activism in Cleveland, 1915–45* (Urbana: University of Illinois Press, 1999); J. Douglas Smith, *Managing White Supremacy: Race, Politics, and Citizenship in Jim Crow Virginia* (Chapel Hill: University of North Carolina Press, 2002). In Matthew C. Whitaker, *Race Work: The Rise of Civil Rights in the Urban West* (Lincoln: University of Nebraska Press, 2005), see especially chapter 2, which focuses on African-American activism in Tuskegee during World War II; Smith, *Managing White Supremacy: Race, Politics, and Citizenship in Jim Crow Virginia*, especially chapter 3; Jennifer E. Brooks, *Defining the Peace: World War II Veterans, Race, and the Remaking of Southern Political Tradition* (Chapel Hill: University of North Carolina Press, 2004); Thomas A. Guglielmo, *White on Arrival: Italians, Race, Color, and Power in Chicago, 1890–1945* (Oxford: Oxford University Press, 2003); Sonya O. Rose, "Gender and Labor History: The Nineteenth-Century Legacy," *International Review of Social History* 38, no. Supplement 1 (1993).

37. Matt García, *A World of its Own: Race, Labor, and Citrus in the Making of Greater Los Angeles, 1900–1970* (Chapel Hill: University of North Carolina Press, 2001), 6; Vicki L. Ruiz, *From Out of the Shadows: Mexican Women in Twentieth-Century America* (Oxford: Oxford University Press, 1998). In his introduction, García offers a succinct summary of the theoretical foundation for including considerations of identity in analyses of political economy. See especially pp. 5–7.
38. Benigno Castillo, "Interview by Jesse J. Escalante," in *Jesse Escalante Oral Histories.*
39. Gilbert Martínez, "Interview by Jesse J. Escalante" in *Jesse Escalante Oral Histories.*
40. Castillo, "Interview by Escalante.
41. Taylor, *Mexican Labor*, 103.
42. Ibid.
43. The building that housed the drugstore also contained three apartments. Although the amount of rent Galindo paid for the location of his store is unknown, it is recorded that he paid $20 a month for his apartment. He lived there with his wife, Juana G., a twenty-one-year-old son who worked as an outdoor track laborer at a steel mill, an eighteen-year-old daughter-in-law, and two small children who were listed as his children, but were probably his grandchildren. The building owner was a Swedish widow who lived with her U.S.-born daughter and English son-in-law. A Mexican couple and two Mexican boarders lived in the third apartment. (1930 U.S. Census, Chicago, IL, ED 2458, p. 5A.
44. Manuel Gamio, "South Chicago: Sr. Galindo," 1926, *Manuel Gamio Papers*. 1930 U.S. Census, Chicago, IL, ED 2458, p. 5A.
45. Lizabeth Cohen, *Making a New Deal: Industrial Workers in Chicago, 1919–1939* (New York: Cambridge University Press, 1990), 110–12.
46. "Los que Réniegan de su Pátria," *Mexico*, Chicago, February 7, 1925.
47. Louise Año Nuevo Kerr, "The Chicano Experience in Chicago: 1920–1970" (Ph.D. diss., University of Illinois at Chicago Circle, 1976), 20, 49–50.
48. "The Mexican Organizations," *Correo Mexicano*, Chicago, September 6, 1926.
49. Ibid.
50. Paul S. Taylor, "List of Societies," 1928, *Paul Schuster Taylor Papers*; Taylor, *Mexican Labor*, 131n70, 133n72; Jones, "Research Notes of Robert Jones," 29.
51. Malachy R. McCarthy, "Which Christ Came to Chicago: Catholic and Protestant Programs to Evangelize, Socialize, and Americanize the Mexican Immigrant, 1900–1940" (Ph.D. diss., Loyola University, 2002), 212–13.
52. Ibid., 221.

53. "The Mexican Mutual Aid Society Benito Juarez," *Heraldo de las Americas*, Chicago, November 1, 1924, trans. by Robert Redfield, in box 59, file 2 of *Robert Redfield Papers.*
54. Martínez, "Interview by Escalante"; Mercedes Rios Radica, "Interview by Jesse J. Escalante," in *Jesse Escalante Oral Histories, Global Communities Collection.* Rios Radica recalled with pride that the consul was present at organizational fundraisers where she would sing and Gilbert Martínez remembers the consul attending performances by a community drama club that was part of a larger South Chicago Mexican recreational organization.
55. Dorothy Anderson, "Book of Sunday Afternoon Mexican Discussion Group, Frente Popular," 1936, in *University of Chicago Settlement Papers*, Chicago History Museum, Chicago. Entry is titled "Monday at Hull House, 2-2-36 3–7pm."
56. Huerta, "Organizaciones Méxicanas."
57. Ibid.
58. "Activities of the Mutual Society Plutarco Elias Calles," *La Lucha*, Chicago, April 21, 1934; "Camp Emiliano Carranza," *El Nacional*, Chicago, April 1, 1931; "Mutualista Obreros Libres Mexicanos," *La Lucha*, Chicago, February 17, 1934.
59. Anita Edgar Jones, "Conditions Surrounding Mexicans in Chicago" (M.A. thesis, University of Chicago, 1928), 167; "Mutualista Obreros Libres Mexicanos"; "To the Members of the Mutualista Independent Mexican Laborers of South Chicago," *El Mexico*, Chicago, October 27, 1928, in *FLPS.*
60. "New Orchestra" *Mexico,* November 7, 1929, in *FLPS.*
61. Ibid.
62. Carlos Perez Lopez, "Interview by Manuel Gamio," Chicago, June 27, 1926, Box 3, File 11, *Manuel Gamio Papers.* His organization, the *Sociedad Fraternal Mexicana,* was founded in 1923 or 1924 and became a lodge of the *Alianza Hispano-Americana* in 1927. Taylor, *Mexican Labor*, 133 n. 72.
63. "What You Ought to Know," *Heraldo de las Americas*, Chicago, November 15, 1924, in Box 59, File 2 in *Robert Redfield Papers.*
64. Jones, "Research Notes of Robert Jones."
65. Rev. Zaldivares, "Interview by Paul S. Taylor," 1928, *Paul Schuster Taylor Papers*; Cheryl R. Ganz and Margaret Strobel, eds., *Pots of Promise: Mexicans and Pottery at Hull House, 1920–40* (Urbana: University of Illinois Press, 2004), 45.
66. Ibid.
67. Paul S. Taylor, "Meeting of Sociedad Mutualistica Ignacio Zaragoza," 1928, *Paul Schuster Taylor Papers.*
68. Between 1933 and 1934, steelworkers in South Chicago and throughout the country turned to the American Federation of Labor (AFL) craft union, the Amalgamated Association of Iron, Steel and Tin Workers (AAISTW), which had represented the steelworkers during the unsuccessful organizing campaign that culminated in the 1919 Great Steel Strike. There is little evidence of Mexican involvement or recruitment of Mexicans by union organizers during this drive. Associated Employees of South Works held its first public meeting in 1935 and affiliated with the newly formed SWOC in 1936. It was not until the affiliation with SWOC, and the targeted recruitment of Mexican workers, that Mexican rank-and-file workers joined the union. For more on the organizing drive of 1933–1934, the creation of Associated Employees, and SWOC, see Lizabeth Cohen, *Making a New Deal*, 293–95.
69. Taylor, *Mexican Labor*, 118–19, 123.
70. Ibid., 121.

71. Alfredo De Avila, "Interview by Jesse J. Escalante," in *Jesse Escalante Oral Histories, Global Communities Collection*; John V. Riffe, "Interview by Nicolas M. Hernandez," December 14, 1936, transcript, *FLPS*.
72. De Avila, "Interview by Escalante."
73. For more on the Mexican involvement in the Steel Workers Organizing Committee (SWOC) and later the United Steel Workers of America (USWA), see Jorge Hernandez-Fujigaki, "Mexican Steelworkers and the United Steelworkers of America in the Midwest: The Inland Steel Experience, 1936–1976," (Ph.D. diss., University of Chicago, 1991). See especially pages 88–109 for involvement before World War II.
74. ". . . And Now Men, to Work," *Correo Mexicano*, Chicago, September 16, 1926.
75. "5,000 Mexicans in Pilsen Park the Night of the 15," *Mexico*, Chicago, September 17, 1927; "And Now Men"; "Festival in South Chicago," *Mexico*, Chicago, September 17, 1927; "Fiestas Patrias organized by the Independent Mexican Commission, A Complete Success: It was the Mexican Workers' Expression of Patriotism," *Mexico*, Chicago, September 25, 1926; "The Independent Mexican Commission Organizes the Program for September 15," *Mexico*, Chicago, August 21, 1926; "What You Ought to Know," *Heraldo de las Americas*.
76. Sánchez, *Becoming Mexican American*, 9–10. For more on created, or "invented," traditions, see Michael Herzfeld, *Ours Once More: Folklore, Ideology, and the Making of Modern Greece* (Austin: University of Texas Press, 1982); Eric Hobsbawm and Terence Ranger, *The Invention of Tradition* (Cambridge: Cambridge University Press, 1983). Also see Jonathan Friedman, "The Past in the Future: History and the Politics of Identity," *American Anthropologist* 94, no. 4 (December 1992); Allan Hanson, "The Making of the Maori: Culture Invention and its Logic," *American Anthropologist* 91, no. 4 (December 1989). Hanson argues that tradition is now commonly understood to be an "invention designed to serve contemporary purposes" to "legitimate or sanctify some current reality or aspiration" (890). The notion of "imagined community" is not limited to Mexican immigrants, of course. Benedict Anderson's exploration of the concept of nationalism—including early models of nationalism created by the Latin American colonial elite's hostility towards Europe—became a staple of scholarship on ethnic identity in the second half of the 1980s and the 1990s. Although Latin Americanists led by Tulio Halperin Donghi have more recently questioned the usefulness for the analysis of colonial Latin America, many continue to use the phrase and concept to describe how heterogeneous groups cohere through the development of extenuated definitions of what constitutes community. For examples, see Nancy P. Appelbaum, *Muddied Waters, Race, Region, and Local History and Colombia, 1846–1948* (Durham, NC: Duke University Press, 2003); and Greg Grandin, "The Instruction of Great Catastrophe: Truth Commissions, National History, and State Formation in Argentina, Chile, and Guatemala," *American Historical Review* 110, no. 1 (February 2005).
77. Taylor, *Mexican Labor*, 138–39.
78. Augustin J. Fink, "Interview by Manuel Gamio," Chicago, June 26, 1926, Box 3, File 11, *Manuel Gamio Papers*; 1930 U.S. Census, Chicago, IL, ED 1711, p. 12B. Augustin Fink Border Crossing Card, Manifests of Statistical and Some Nonstatistical Alien Arrivals at Laredo, Texas, May 1903–April 1955, Record Group: 85, Records of the Immigration and Naturalization Service, Microfilm Serial A3437, Microfilm Roll 77.
79. Perez Lopez, "Interview by Manuel Gamio."
80. Ibid.
81. Juan B. Medina "Interview by Manuel Gamio," June 21, 1926, Transcript, Box 3, File 11, *Manuel Gamio Papers*; 1930 U.S. Census, Chicago, IL, ED 1642, p. 10A.

82. Guzman, "Interview by Escalante."
83. Martínez, "Interview by Escalante."

CHAPTER 7

1. Jose Cruz Diaz, "Interview by Jesse J. Escalante," in *Jesse Escalante Oral Histories, Global Communities Collection,* Chicago History Museum.
2. Mercedes Rios Radica, "Interview by Escalante," in *Jesse Escalante Oral Histories.*
3. Ibid.
4. Cruz Diaz, "Interview by Escalante"; Agapita Flores, "Interview by Escalante," in *Jesse Escalante Oral Histories.*
5. Rios Radica, "Interview by Escalante."
6. Alfredo De Avila,"Interview by Jesse J. Escalante," in *Jesse Escalante Oral Histories*; Rios Radica, "Interview by Escalante." One of the "Mexican" jobs included laying and moving railroad tracks closer to the lakeshore in order to dump the furnace slag into the lake. The job required the constant movement of the railroad tracks as the shoreline filled with slag.
7. Cruz Diaz, "Interview by Escalante."
8. Louise Año Nuevo Kerr, "The Chicano Experience in Chicago: 1920–1970" (Ph.D. diss., University of Illinois at Chicago Circle, 1976), 75.
9. Max Guzman, "Interview by Escalante," in *Jesse Escalante Oral Histories.*
10. "Acting Mexican Consul Jailed," *El Nacional,* Chicago, July 11, 1931.
11. "Mexico Asks for Probe of Term Given Consul: Agent Freed on Serving Two Hours in Chicago Contempt Case," *Washington Post,* July 8, 1931.
12. Ibid.
13. The United Press reporter quotes Green in a widely distributed newswire story. My source is "Mexicans Aroused by Case of Consul," *Charleston Daily Mail,* July 9, 1931.
14. "Washington Makes Apology to Mexico: Regret at Sentencing Consul in Chicago," *New York Times,* July 11, 1931; "Governor to Probe Mexican's Sentence," *Atlanta Constitution,* July 9, 1931; "Mexico Asks for Probe"; "Protest by Mexico Frees Jailed Consul: Sentenced and Freed," *New York Times,* July 10, 1931; 1900 United States Census, Chicago, IL, ED 1067, p. 17B.
15. "Washington Makes Apology"; "Governor to Probe Mexican's Sentence"; "Mexico Asks for Probe"; "Protest by Mexico."
16. Gabriela F. Arredondo, "Navigating Ethno-Racial Currents: Mexicans in Chicago, 1919–1939," *Journal of Urban History* 30, no. 3 (2004): 410.
17. The classic work on forced repatriation during the Great Depression is Abraham Hoffman, *Unwanted Mexican Americans in the Great Depression: Repatriation Pressures, 1926–1939* (Tucson: University of Arizona Press, 1974). His focus, like that of most scholars of repatriation, is on the organized repatriation programs in the Los Angeles area. Hoffman, in "Mexican Repatriation Statistics: Some Suggested Alternatives to Carey McWilliams," *Western Historical Quarterly* 3, no. 4 (October 1972) 391–404, provides a critical analysis of earlier numbers. For forced repatriation in the American Midwest, see especially Juan R. García, *Mexicans in the Midwest, 1900–1932* (Tucson: University of Arizona Press, 1996), 231–35. See also Mark Reisler, *By the Sweat of Their Brow: Mexican Immigrant Labor in the United States, 1900–1940* (Westport, CT: Greenwood Press, 1976), 231–32; and Zaragosa Vargas, *Proletarians of the North: A History of Mexican Industrial Workers in Detroit and the Midwest, 1917–1933* (Berkeley: University of California Press, 1993), 189–90.
18. Kerr, "The Chicano Experience," 75.

19. F. C. Harrington, "Questions and Answers on the W.P.A." (Washington, DC: U.S. Government Printing Office, 1939).
20. Georgia Elma Harkness, *The Church and the Immigrant* (New York: George H. Doran Company, 1921), 43.
21. Harrington, "W.P.A. Pamphlet."
22. The Committee on the Long-Range Work and Relief Policies, "Security, Work, and Relief Policies: Report of the Committee on Long-Range Work and Relief Policies to the National Resources Planning Board" (Washington, DC: Government Printing Office, 1942), 237.
23. Patrick W. Ettinger, *Imaginary Lines: Border Enforcement and the Origins of Undocumented Immigrants, 1882–1930* (Austin: University of Texas Press, 2009), 146.
24. De Avila, "Interview by Escalante"; Lucio Franco, "Interview by Jesse J. Escalante," in *Jesse Escalante Oral Histories.*
25. Francisco Balderama and Raymond Rodríguez, *Decade of Betrayal: Mexican Repatriation in the 1930s* (Albuquerque: University of New Mexico Press, 1995), 56–57.
26. Dionicio Nodín Valdés, *Barrios Norteños: St. Paul and Midwestern Mexican Communities in the Twentieth Century* (Austin: University of Texas Press, 2000), 96–97.
27. David Gutiérrez, *Walls and Mirrors: Mexican Americans, Mexican Immigrants, and the Politics of Ethnicity* (Berkeley: University of California Press, 1995), 72–74.
28. Hoffman, *Unwanted Mexican Americans*, 126. On page 166, Hoffman defines repatriation as follows: "A. Voluntary repatriation by the aliens themselves, B. Repatriation of destitute aliens by the federal government, C. Organized repatriation by local private and public welfare agencies, D. Organized repatriation of aliens by the Mexican consul and the Mexican ethnic community, E. 'Coercive,' 'forced,' or 'involuntary' repatriation (quotation marks indicate that the repatriation was not actually desired by the alien, but was forced upon him)." Mexicans in South Chicago experienced and participated in all of these forms of repatriation.
29. U.S. Bureau of the Census, *Abstract of the Fifteenth Census*, 98; U.S. Bureau of the Census, *Sixteenth Census of the United States: 1940, Population* (Washington, DC: U.S. Government Printing Office, 1943), 33–34; Chicago Council of Social Agencies, *Social Service Directory and Yearbook* (Chicago: Chicago Council of Social Agencies, 1933), 199–229; Ernest W. Burgess and Charles Newcomb, eds., *Census Data of the City of Chicago, 1930* (Chicago: University of Chicago Press, 1933), xi–xiii; Eunice Felter, "The Social Adaptations of the Mexican Churches in the Chicago Area" (M.A. thesis, University of Chicago, 1941), 11; Charles Newcomb and Richard O. Lang, eds., *Census Data of the City of Chicago, 1934* (Chicago: University of Chicago Press, 1934).
30. Edward Jackson Baur, "Delinquency Among Mexican Boys in South Chicago" (M.A. thesis, University of Chicago, 1938), 27, 45; Eunice Felter, "The Social Adaptations of the Mexican Churches in the Chicago Area" (M.A. thesis, University of Chicago, 1941), 11.
31. Año Nuevo Kerr, "Chicano Experience," 75.
32. Francisco Arturo Rosales and Daniel T. Simon, "Mexican Immigrant Experience in the Urban Midwest: East Chicago, Indiana, 1919–1945," in *Forging a Community: the Latino Experience in Northwest Indiana, 1919–1975*, eds. James B. Lane and Edward J. Escobar (Chicago: Cattails Press, 1987), 146–47.
33. Hoffman, *Unwanted Mexican Americans*, ix. Although Hoffman's institutional history discusses nationwide repatriation programs and statistics, his focus is on Los Angeles.
34. Gutiérrez, *Walls and Mirrors*, 72–74.
35. Hoffman, *Unwanted Mexican Americans*, ix.

36. Ibid, 3.
37. Rosales and Simon, "Mexican Immigrant Experience," 147–49.
38. Ciro Sepúlveda, "La Colonia Del Harbor: A History of Mexicanos in East Chicago, Indiana, 1919–1932" (Ph.D. diss., Notre Dame University, 1976), 142–43; Francisco Arturo Rosales, "Mexican Immigration to the Urban Midwest during the 1920s" (Ph.D. diss., Indiana University, 1978), 231.
39. Raymond A. Mohl and Neil Betten, "Discrimination and Repatriation: Mexican Life in Gary," in *Forging a Community: The Latino Experience in Northwest Indiana, 1919–1975*, eds. James B. Lane and Edward J. Escobar (Chicago: Cattails Press, 1987), 166.
40. Ibid, 173, 176–77.
41. Valdés, *Barrios Norteños*, 97–98.
42. "Deport 200 in Chicago War on Smuggling Ring," *Chicago Tribune*, October 27, 1931.
43. Ibid.
44. Ibid.
45. *Los Angeles Times*, April 11, 1931, in F. Arturo Rosales, ed., *Testimonio: A Documentary History of the Mexican American Struggle for Civil Rights* (Houston: Arte Público Press, 2000), 96–97. In the article, Doak finds it necessary to point out he was not targeting "Communists or any other special class" in the deportation drives.
46. This concept of immigrants being responsible for unemployment at times of economic crisis is the official justification for the restrictive Alabama anti-immigrant laws of 2011 and 2012.
47. Hoffman, *Unwanted Mexican Americans*, 117.
48. Ibid., 116–17; Taylor, *Mexican Labor*, 48.
49. Arnoldo de León, *Ethnicity in the Sunbelt: A History of Mexican Americans in Houston* (Houston: Mexican American Studies Program, University of Houston, 1989), 47–48.
50. Ibid. For more on repatriation in cities other than Chicago, see de León, *Ethnicity in the Sunbelt;* Hoffman, *Unwanted Mexican Americans;* George J. Sánchez, *Becoming Mexican American: Ethnicity, Culture and Identity in Chicano Los Angeles, 1900–1945* (New York: Oxford University Press, 1993); Sepúlveda, "La Colonia del Harbor"; Valdés, *Barrios Norteños*, 97–98. For more on earlier Mexican government efforts in encouraging repatriation of Mexicans from throughout the United States and in some cases financing transportation costs, see Fernando Saúl Alanis Enciso, "No cuenten conmigo: La pólitica de repatriación del gobierno mexicano y sus nacionales en Estados Unidos, 1910–1928," *Mexican Studies/Estudios Mexicanos* 19, no. 2 (Summer 2003).
51. De León, *Ethnicity in the Sunbelt*, 48; Gutiérrez, *Walls and Mirrors*, 72–74.
52. Rios Radica, "Interview by Escalante."
53. "Letter from Consul," *El Nacional*, Chicago, May 14, 1932.
54. Ibid. "Return of Mexicans to Homeland," *El Nacional*, Chicago, May 28, 1932.
55. Rios Radica, "Interview by Escalante."
56. Ibid.
57. De Avila, "Interview by Escalante."
58. Serafín García, "Interview by Jesse J. Escalante," in *Jesse Escalante Oral Histories*.
59. "A Repatriation Program," *La Defensa*, Chicago, January 18, 1936.
60. Antonio L. Schmidt, "Interview by Victor Chavez," May 19, 1937, *FLPS*.
61. Dorothy Anderson, "Book of Sunday Afternoon Mexican Discussion Group, Frente Popular," in *University of Chicago Settlement House Papers*, Chicago Historical Society, Chicago.
62. Ibid.

63. Ibid.
64. Ibid. For more on repatriation, see Hoffman, *Unwanted Mexican Americans*, and Rosales and Simon, "Mexican Immigrant Experience."
65. John Henry Flores, "On the Wings of the Revolution: Transnational Politics and the Making of Mexican American Identities" (Ph.D. diss., University of Illinois at Chicago, 2009), 89–90. See chapter 3 of Flores for an in-depth discussion of the *Frente Popular* in Chicago.
66. "Mexicans on Relief in Chicago—October 1933," in box 52, file 2, *Ernest W. Burgess Papers*, University of Chicago Special Collections, Chicago.
67. Jesse John Escalante, "History of the Mexican Community in South Chicago" (M.A. thesis, Northeastern Illinois University, 1982), 20–21; Jean Rodríguez, interview by Jesse Escalante, January 22, 1980, in Escalante, "History of the Mexican Community in South Chicago," 21.
68. Escalante, "Mexican South Chicago," 19–20.
69. Rafael Guardado, "Interview by Jesse J. Escalante," in *Jesse Escalante Oral Histories*; Rios Radica, "Interview by Escalante"; Anthony Romo, "Interview by Jesse J. Escalante," in *Jesse Escalante Oral Histories.*
70. Rios Radica, "Interview by Escalante."
71. "There will be Elections in the Blue Cross of South Chicago," *El Nacional*, Chicago, June 20, 1931.
72. Ibid.
73. Dr. Oscar G. Carrera, "A Message to the Mexican Blue Cross," *Mexico*, Chicago, April 12, 1930; Narciso González, "The Problem of the Mexican Blue Cross of Chicago," *Mexico*, Chicago, April 5, 1930; "Mexican Blue Cross of South Chicago," *El Nacional*, Chicago, December 17, 1930.
74. González, "The Problem of the Mexican Blue Cross of Chicago."
75. Ibid.
76. "A Scene of Tragedy," *El Nacional*, May 14, 1932, in *FLPS.*
77. Ibid.
78. Ibid.
79. Ibid.
80. Carrera, "A Message to the Mexican Blue Cross."
81. "Mexican Blue Cross Forms Auxiliary in South Chicago," *Mexico*, Chicago, May 6, 1930; "Mexican Blue Cross of South Chicago"; "Blue Cross Elections." The officer positions for the South Chicago Auxiliary of the *Cruz Azul Mexicana* included: president, vice president, secretary, treasurer, director of publicity, director of nurses, director of ambulances, director of festivities, counsel, and flag bearer.
82. Anderson, "Frente Popular Notebook."
83. Ibid.

CHAPTER 8

1. Manuel Bravo, "Interview by Jesse J. Escalante," in *Jesse Escalante Oral Histories, Global Communities Collection,* Chicago History Museum; Serafín García, "Interview by Jesse J. Escalante," in *Jesse Escalante Oral Histories.*
2. Bravo, "Interview by Escalante"; García, "Interview by Escalante."
3. Susan Currell, *The March of Spare Time: The Problem and Promise of Leisure in the Great Depression* (Philadelphia: University of Pennsylvania Press, 2005), 2–3.
4. Jeffrey Hill, *Sport, Leisure, and Culture in Twentieth-Century Britain* (New York: Palgrave, 2002), 6.

5. See Allen Guttmann, "Sport, Politics and the Engaged Historian," *Journal of Contemporary History* 38, no. 3 (July 2003), for a helpful discussion of politics, sports, and the role of historians in linking the two.
6. José Alamillo, *Making Lemonade out of Lemons: Mexican American Labor and Leisure in a California Town, 1880–1960* (Urbana: University of Illinois Press, 2006), 65.
7. Robert C. Jones and Louis R. Wilson, *The Mexican in Chicago, The Racial and Nationality Groups of Chicago: Their Religious Faiths and Conditions* (Chicago: Comity Commission of the Chicago Church Federation, 1931). These quotations were found in a summary transcription of the report located in the *FLPS*.
8. Alamillo, *Making Lemonade*, 64.
9. For more on South Chicago pool rooms, see Gabriela Arredondo, *Mexican Chicago: Race, Identity, and Nation, 1916–39* (Urbana: University of Illinois Press, 2008), 22, 53. One example of police harassment of pool hall patrons is detailed in "Impudence of Policemen," *El Nacional*, January 7, 1931.
10. Max Guzman, "Interview by Jesse J. Escalante," in *Jesse Escalante Oral Histories*; Anthony Romo, "Interview by Jesse J. Escalante," in *Jesse Escalante Oral Histories;* Anita Edgar Jones, "Conditions Surrounding Mexicans in Chicago" (M.A. thesis, University of Chicago, 1928), 72–73; Jesse John Escalante, "History of the Mexican Community in South Chicago" (M.A. thesis, Northeastern Illinois University, 1982), 23; Sidney Levin, "Interview by Jesse J. Escalante," in *Jesse Escalante Oral Histories.*
11. Marian Lorena Osborn, "The Development of Recreation in the South Park System of Chicago" (M.A. thesis, University of Chicago, 1928), 1–2.
12. Levin, "Interview by Escalante"; Romo, "Interview by Escalante"; Jones, "Conditions Surrounding Mexicans," 72–73; Escalante, "Mexican South Chicago," 23; Martínez, "Interview by Escalante." Chicago-area Mexican youth frequently referred to working-class European immigrants (other than the Irish and Italian) as Polish regardless of their national origin.
13. Guzman, "Interview by Escalante"; *Leisure Time Directory* (Chicago, 1940), 133–34. The City of Chicago listed Russell Square Park and Rocky Ledge Park as the other two public parks within the South Chicago neighborhood.
14. Jones, "Conditions," 72.
15. Bird Memorial (9135 S. Brandon Avenue) was located only one-half of a block from Our Lady of Guadalupe (northeast corner of 91st Street and South Brandon Avenue). Our Lady of Guadalupe's church building dominates that part of the neighborhood; Bird Memorial was razed.
16. Levin, "Interview by Escalante"; Martínez, "Interview by Escalante"; *Leisure Time Directory;* Arthur J. Todd, William F. Byron, and Howard L. Vierow, *The Chicago Recreation Survey: Private Recreation*, 5 vols., vol. III (Chicago: Chicago Recreation Commission and Northwestern University, 1937), 27. Henry Seymour Brown et al., "Report of the Committee on Findings set up by the Comity Commission of the Chicago Federation of Churches on the Survey of South Chicago by the Department of Research and Survey under the Direction of Professor Holt" (1928), Comity Committee Folder, box 9 (new), *Chicago Church Federation Papers*, Chicago History Museum, 3.
17. Jose Cruz Diaz, "Interview by Jesse J. Escalante," in *Jesse Escalante Oral Histories*. For more on the struggle between the Roman Catholic church and Protestant churches over the "hearts and minds" of Mexican Chicagoans, see Anne M. Martínez, "Bordering on the Sacred: Religion, Nation, and U.S.-Mexican Relations, 1910–1929" (Ph.D. diss., University

of Minnesota, 2003), and Malachy Richard McCarthy, "Which Christ Came to Chicago: Catholic and Protestant Programs to Evangelize, Socialize, and Americanize the Mexican Immigrant, 1900–1940" (Ph.D. diss., Loyola University, 2002).

18. Raymond Sanford, "Interview by Victor Chavez," December 8, 1936, *FLPS*; Todd, et al., *Chicago Recreation Survey, Vol. III*, 158. The Recreational Survey states that Common Ground served twenty-six nationalities with members of the Polish community predominating.
19. Sanford, "Interview by Chavez"; Todd, et al., *Chicago Recreation Survey, Vol. III*, 158.
20. Cruz Diaz, "Interview by Escalante"; Sanford, "Interview by Chavez."
21. Mary E. Odem, *Delinquent Daughters: Protecting and Policing Adolescent Female Sexuality in the United States, 1885–1920* (Chapel Hill: University of North Carolina Press, 1995), 1. For the experiences of ethnic European women immigrants in Chicago, see Joanne J. Meyerowitz, *Women Adrift: Independent Wage Earners in Chicago, 1880–1930* (Chicago: University of Chicago Press, 1988). Directories of general social services and recreational facilities list several Protestant-run programs and sites; Catholic programs specifically for Mexicans did not appear in South Chicago until the late 1930s. Louise Año Nuevo Kerr, "The Chicano Experience in Chicago: 1920–1970" (Ph.D. diss., University of Illinois at Chicago Circle, 1976), 57–8; Brown et al., "Comity Report; City of Chicago," *Leisure Time Directory*; Todd, et al., *Chicago Recreation Survey, Vol III.*
22. Socorro Zaragoza, interview by Paul S. Taylor, box 11, file 32, *Paul Schuster Taylor Papers*, BANC MSS 84/38 c, The Bancroft Library, University of California, Berkeley; Arredondo, *Mexican Chicago: Race, Identity, and Nation, 1916–39*, 119–25; Vicki Ruiz, "'Star Struck': Acculturation, Adolescence, and the Mexican American Woman, 1920–1950," in *Building With Our Hands: New Directions in Chicana Studies*, eds. Adela de la Torre and Beatriz Pesquera (Berkeley: University of California Press, 1993).
23. Edward Jackson Baur, "Delinquency Among Mexican Boys in South Chicago" (M.A. thesis, University of Chicago, 1938), 55, 98; Levin, "Interview by Escalante"; Vicki L. Ruiz, *From out of the Shadows: Mexican Women in Twentieth-Century America* (Oxford: Oxford University Press, 1998). Ruiz provides a thorough examination of the conflicts between Mexican parents and their daughters in the Southwest caused by the girls' "Americanization," particularly in social activities.
24. Cruz Diaz, "Interview by Escalante"; Romo, "Interview by Escalante"; Escalante, "Mexican South Chicago," 28; Baur, "Delinquency," 171.
25. Escalante, "Mexican South Chicago," 28; Romo, "Interview by Escalante"; Baur, "Delinquency," 132, 134.
26. "Youth Basketball," *El Nacional*, Chicago, February 17, 1934.
27. Baur, "Delinquency," 152; Martínez, "Interview by Escalante"; Cruz Diaz, "Interview by Escalante."
28. Romo, "Interview by Escalante."
29. Martínez, "Interview by Escalante"; Baur, "Delinquency," 152.
30. "The New Basket Ball Team," *El Nacional*, January 9, 1932, in *FLPS*.
31. "Atlas Basket Ball Team in Action," *El Nacional*, March 12, 1932.
32. Ibid.
33. Ibid.
34. Cruz Diaz, "Interview by Escalante"; Levin, "Interview by Escalante"; Pete Martínez, "Interview by Jesse J. Escalante," in *Jesse Escalante Oral Histories*.

35. Indoor ball was a precursor to softball. It was, in essence, baseball played indoors with a ball akin to a softball. Official games lasted seven innings.
36. Justino Cordero, "Interview by Jesse J. Escalante," in *Jesse Escalante Oral Histories*; de Avila, "Interview by Escalante"; Levin, "Interview by Escalante"; Eduardo Peralta, "Interview by Nicolas M. Hernandez," Chicago, December 11, 1936, *FLPS*.
37. Peralta, "Interview by Hernandez."
38. Alberto Cuellar, "Interview by Nicolas M. Hernandez," 1936, *FLPS*.
39. Pete Martínez, "Interview by Escalante."
40. Escalante, "Mexican South Chicago," 24.
41. Pete Martínez, "Interview by Escalante"; Cruz Diaz, "Interview by Escalante"; García, "Interview by Escalante"; Escalante, "Mexican South Chicago," 24; Baur, "Delinquency," 132.
42. Gary Ross Mormino, "The Playing Fields of St. Louis: Italian Immigrants and Sports, 1925–1941," *Journal of Sports History* 9, no. 2 (Summer 1982): 5.
43. Ibid.
44. Alamillo, *Making Lemonade*, 107.
45. Douglas Monroy, *Rebirth: Mexican Los Angeles from the Great Migration to the Great Depression* (Berkeley: University of California Press, 1999), 46.
46. See Steven A. Riess, ed., *Sports and the American Jew* (Syracuse: Syracuse University Press, 1998), for a collection of essays.
47. Martínez, "Interview by Escalante"; Cruz Diaz, "Interview by Escalante"; "Monterrey Baseball Team Defeats Drexel Square A.C.," *El Nacional*, Chicago, September 2, 1933; Baur, "Delinquency," 152; Francisco Arturo Rosales and Daniel T. Simon, "Mexican Immigrant Experience in the Urban Midwest: East Chicago, Indiana, 1919–1945," *Indiana Magazine of History* 77, no. 4 (December 1981): 340.
48. Levin, "Interview by Escalante."
49. García, "Interview by Escalante"; Martínez, "Interview by Escalante"; Levin, "Interview by Escalante"; Bravo, "Interview by Escalante"; "Cuauhtemoc Club Festival," *El Nacional*, Chicago, January 6, 1934.

EPILOGUE

1. Mike Amezcua, "The Second City Anew: Mexicans, Urban Culture, and Migration in the Transformation of Chicago, 1940–1965" (Ph.D. diss., Yale University, 2011); Lilia Fernández, *Brown in the Windy City: Mexicans and Puerto Ricans in Postwar Chicago* (Chicago: University of Chicago Press, 2012).

BIBLIOGRAPHY

ARCHIVES

Bancroft Library, University of California-Berkeley
  Paul S. Taylor Papers
  Manuel Gamio Papers
Chicago History Museum
  Foreign Populations Collection
  Chicago Church Federation Papers
  Jesse Escalante Oral Histories, Global Communities Collection
  University of Chicago Settlement House Papers
Chicago Theological Seminary
  Mexicans in Chicago Archival Collection.
Claretian Missionaries Archives USA
  Our Lady of Guadalupe Parish Collection
Southeast Chicago Historical Society
  Photographic Collection
University of Chicago
  E. W. Burgess Papers
  Chicago Foreign Language Press Survey
  Robert Redfield Papers
  Louis Wirth Papers
University of Illinois at Chicago
  Immigrant Protective League Papers
  Hull House Papers

NEWSPAPERS

Atlanta Constitution
Chicago Defender
Chicago Tribune
Correo Mexicano, Chicago
La Defensa, Chicago
El Heraldo de Las Americas, Chicago
La Lucha, Chicago
Mexico, Chicago
El Nacional, Chicago

Secondary Sources

Acuña, Rodolfo. *A Community under Siege: A Chronicle of Chicanos East of the Los Angeles River, 1945–1975*. Los Angeles: Chicano Studies Research Center, Publications, University of California at Los Angeles, 1984.

———. *Occupied America: A History of Chicanos*. 3rd ed. New York: Harper & Row, 1988.

Adams, Mary Faith. "Present Housing Conditions in South Chicago, South Deering and Pullman." M.A. thesis, University of Chicago, 1926.

Aguilar Camín, Héctor, and Lorenzo Meyer. *In the Shadow of the Mexican Revolution: Contemporary Mexican History, 1910-1989*. Translations from Latin America Series. Austin: University of Texas Press, 1993.

Alamillo, José. *Making Lemonade out of Lemons: Mexican American Labor and Leisure in a California Town, 1880–1960*. Urbana: University of Illinois Press, 2006.

Almaguer, Tomás. *Racial Fault Lines: The Historical Origins of White Supremacy in California*. Berkeley: University of California Press, 1994.

Alter, Peter Thomas. "Mexicans and Serbs in Southeast Chicago: Racial Group Formation During the Twentieth Century." *Journal of the Illinois State Historical Society* 94, no. 4 (2001): 403–19.

———. "The Serbian Great Migration : Serbs in the Chicago Region, 1880s to 1930s." Ph.D. diss., University of Arizona, 2000.

Amezcua, Mike. "The Second City Anew: Mexicans, Urban Culture, and Migration in the Transformation of Chicago, 1940–1965." Ph.D. diss., Yale University, 2011.

Appleton, John B. "The Iron and Steel Industry in the Calumet District: A Study in Economic Geography." *University of Illinois Studies in the Social Sciences* XIII, no. 2 (1925): 11–131.

Arredondo, Gabriela F. *Mexican Chicago: Race, Identity, and Nation, 1916–39*. Urbana: University of Illinois Press, 2008.

———. "Navigating Ethno-Racial Currents." *Journal of Urban History* 30, no. 3 (2004): 399–427.

Audoin-Rouzeau, Stéphane, and Annette Becker. *14–18, Understanding the Great War*. New York: Hill and Wang, 2002.

Bada, Xóchitl. "Mexican Hometown Associations in Chicago: The Newest Agents of Civic Participation." In *¡Marcha!: Latino Chicago and the Immigrant Rights Movement*, edited by Amalia Pallares and Nilda Flores-González, 146–62. Urbana: University of Illinois Press, 2010.

Badillo, David A. *Latinos in Michigan*. East Lansing: Michigan State University Press, 2003.

Balderrama, Francisco E., and Raymond Rodriguez. *Decade of Betrayal: Mexican Repatriation in the 1930s*. Revised ed. Albuquerque: University of New Mexico Press, 2006.

Baron, Ava. *Work Engendered: Toward a New History of American Labor*. Ithaca: Cornell University Press, 1991.

Barrett, James R. *Work and Community in the Jungle: Chicago's Packinghouse Workers, 1894–1922*. Urbana: University of Illinois Press, 1987.

Baur, Edward Jackson. "Delinquency among Mexican Boys in South Chicago." M.A. thesis, University of Chicago, 1938.

Bender, Thomas. *Community and Social Change in America*. New Brunswick: Rutgers University Press, 1978.

Black, Edwin. *War against the Weak: Eugenics and America's Campaign to Create a Master Race*. Expanded Edition. New York: Dialog Press, 2012.

Blea, Irene I. *La Chicana and the Intersection of Race, Class, and Gender*. New York: Praeger, 1992.
Bodnar, John E. *The Transplanted: A History of Immigrants in Urban America*. Bloomington: Indiana University Press, 1985.
Breckinridge, Sophonisba P., and Edith Abbott. "Housing Conditions in Chicago, III: Back of the Yards." *American Journal of Sociology* 16, no. 4 (January 1911): 433–68.
———. "Housing Conditions in Chicago, IV: The West Side Revisited." *American Journal of Sociology* 17, no. 1 (July 1911): 1–34.
Broder, Sherri. *Tramps, Unfit Mothers, and Neglected Children: Negotiating the Family in Nineteenth-Century Philadelphia*. Philadelphia: University of Pennsylvania Press, 2002.
Brody, David. *Labor in Crisis: The Steel Strike of 1919*. Philadelphia: Lippincott, 1965.
Burgess, E. W., and Charles Shelton Newcomb. *Census Data of the City of Chicago, 1920*. Chicago: University of Chicago Press, 1931.
———, ———, and University of Chicago Social Science Research Committee. *Census Data of the City of Chicago, 1930*. Chicago: University of Chicago Press, 1933.
Calhoun, Craig J. *Habermas and the Public Sphere*. Studies in Contemporary German Social Thought. Cambridge: MIT Press, 1992.
Camarillo, Albert. *Chicanos in a Changing Society: From Mexican Pueblos to American Barrios in Santa Barbara and Southern California, 1848–1930*. Cambridge: Harvard University Press, 1979.
Camblon, Ruth S. "Mexicans in Chicago," *The Family* VII, no. 7, 1926.
Carr, Barry. *El Movimiento Obrero Y La Política En México, 1910–1929*. 2 vols, Sepsetentas. México: Secretaría de Educación Pública, Dirección General de Divulgación, 1976.
Chicago Recreation Commission. *The Chicago Recreation Survey, 1937*. 5 vols. Chicago: Chicago Recreation Commission, 1937.
———. *Leisure Time Directory: Chicago Public and Semi-Public Recreation and Auxiliary Agencies*. Chicago: Chicago Recreation Commission, 1936.
——— *Leisure Time Directory*. Chicago: Chicago Recreation Commission, 1937.
Cohen, Lizabeth. *Making a New Deal: Industrial Workers in Chicago, 1919–1939*. New York: Cambridge University Press, 1990.
Corwin, Arthur F. *Immigrants—and Immigrants: Perspectives on Mexican Labor Migration to the United States*. Westport, CT: Greenwood Press, 1978.
Cressey, Paul Frederick. "The Succession of Cultural Groups in the City of Chicago." Ph.D. diss. , University of Chicago, 1930.
Cressey, Paul Goalby. *The Taxi-Dance Hall: A Sociological Study in Commercialized Recreation and City Life*. Chicago: University of Chicago Press, 1932.
Cronon, William. *Nature's Metropolis: Chicago and the Great West*. New York: W.W. Norton, 1991.
Cunningham, William James. *American Railroads: Government Control and Reconstruction Policies*. Chicago: A.W. Shaw, 1922.
Currell, Susan *The March of Spare Time: The Problem and Promise of Leisure in the Great Depression*. Philadelphia: University of Pennsylvania Press, 2005.
De León, Arnoldo. *Ethnicity in the Sunbelt: Mexican Americans in Houston*. College Station: Texas A&M University Press, 2001.
Deutsch, Sarah. *No Separate Refuge: Culture, Class, and Gender on an Anglo-Hispanic Frontier, 1880–1940*. New York: Oxford University Press, 1987.

Drolet, Godias J. "Discussion of Tuberculosis in Mexicans." *American Journal of Public Health and the Nation's Health* 9, no. 3 (March 1929): 285–86.

DuBois, Ellen Carol, and Vicki Ruíz. *Unequal Sisters: A Multicultural Reader in U.S. Women's History*. 2nd ed. New York: Routledge, 1994.

Ettinger, Patrick W. *Imaginary Lines: Border Enforcement and the Origins of Undocumented, 1882–1930*. Austin: University of Texas Press, 2009.

Etzioni, Amitai, and Jared Bloom. *We Are What We Celebrate: Understanding Holidays and Rituals*. New York: New York University Press, 2004.

Ewen, Elizabeth. *Immigrant Women in the Land of Dollars: Life and Culture on the Lower East Side, 1890–1925*. New York: Monthly Review Press, 1985.

Felter, Eunice. "The Social Adaptations of the Mexican Churches in the Chicago Area." M.A. thesis, University of Chicago, 1941.

Flores, John Henry. "On the Wings of the Revolution: Transnational Politics and the Making of Mexican American Identities." Ph.D. diss., University of Illinois at Chicago, 2009.

Foley, Neil. *The White Scourge: Mexicans, Blacks, and Poor Whites in Texas Cotton Culture*. Berkeley: University of California Press, 1997.

Foster, William Z. *The Great Steel Strike and Its Lessons*. New York: B. W. Huebsch, 1920.

Fouad, Nadya A., and Patricia M. Arredondo. *Becoming Culturally Oriented: Practical Advice for Psychologists and Educators*. Washington, DC: American Psychological Association, 2007.

Gabaccia, Donna R. *From the Other Side: Women, Gender, and Immigrant Life in the U.S., 1820–1990*. Bloomington: Indiana University Press, 1994.

Gamio, Manuel. *The Mexican Immigrant: His Life-Story*. Chicago: University of Chicago Press, 1931.

———. *Mexican Immigration to the United States; a Study of Human Migration and Adjustment*. Chicago: University of Chicago Press, 1930.

———, Devra Weber, Roberto Melville, and Juan Vicente Palerm. *El Inmigrante Mexicano: La Historia De Su Vida : Entrevistas Completas, 1926–1927*. México: University of California CIESAS, 2002.

———, and Gilberto Loyo. *El Inmigrante Mexicano; La Historia De Su Vida*. México: Universidad Nacional Autónoma de México, 1969.

Ganz, Cheryl R., and Margaret Strobel, eds. *Pots of Promise: Mexicans and Pottery at Hull House, 1920–40*. Urbana: University of Illinois, 2004.

García, Juan R. *Mexicans in the Midwest, 1900–1932*. Tucson: University of Arizona Press, 1996.

García, Mario T. *Desert Immigrants : The Mexicans of El Paso, 1880–1920*. New Haven: Yale University Press, 1981.

García, Richard A. *Rise of the Mexican American Middle Class: San Antonio, 1929–1941*. College Station: Texas A&M University Press, 1991.

Garcilazo, Jeffrey Marcos. "'Traqueros': Mexican Railroad Workers in the United States, 1870–1930." Ph.D. diss., University of California, Santa Barbara, 1995.

Gilly, Adolfo. *La Revolución Interrumpida; México, 1910–1920: Una Guerra Campesina Por La Tierra Y El Poder*. México: Ediciones El Caballito, 1971.

Glenn, Susan Anita. *Daughters of the Shtetl: Life and Labor in the Immigrant Generation*. Ithaca: Cornell University Press, 1990.

Goldberg, Benjamin J. "Tuberculosis in Racial Types with Special Reference to Mexicans." *American Journal of Public Health and the Nation's Health* 9, no. 3 (March 1929): 274–84.

Gómez-Quiñones, Juan. *Chicano Politics: Reality and Promise, 1940–1990*. Albuquerque: University of New Mexico Press, 1990.

———. *Mexican American Labor, 1790–1990*. Albuquerque: University of New Mexico Press, 1994.

González, Gilbert G. *Mexican Consuls and Labor Organizing: Imperial Politics in the American Southwest*. Austin: University of Texas Press, 1999.

Griswold del Castillo, Richard. *The Los Angeles Barrio, 1850–1890: A Social History*. Berkeley: University of California Press, 1979.

Grossman, James R. *Land of Hope: Chicago, Black Southerners, and the Great Migration*. Chicago: University of Chicago Press, 1989.

———, Ann Durkin Keating, Janice L. Reiff, Newberry Library, and Chicago Historical Society. *The Encyclopedia of Chicago*. Chicago: University of Chicago Press, 2004.

Guérin-Gonzales, Camille. *Mexican Workers and American Dreams: Immigration, Repatriation, and California Farm Labor, 1900–1939*. New Brunswick: Rutgers University Press, 1994.

Guglielmo, Thomas A. *White on Arrival: Italians, Race, Color, and Power in Chicago, 1890–1945*. New York: Oxford University Press, 2003.

Gutierrez, David. *Between Two Worlds: Mexican Immigrants in the United States*, Jaguar Books on Latin America. Wilmington, DE.: Scholarly Resources, 1996.

———. *Walls and Mirrors: Mexican Americans, Mexican Immigrants, and the Politics of Ethnicity*. Berkeley: University of California Press, 1995.

Gutman, Herbert George. *Work, Culture, and Society in Industrializing America: Essays in American Working-Class and Social History*. New York: Vintage Books, 1977.

Guttmann, Allen. "Sport, Politics and the Engaged Historian." *Journal of Contemporary History* 38, no. 3 (July 2003): 363–75.

Halpern, Rick. *Down on the Killing Floor: Black and White Workers in Chicago's Packinghouses, 1904–54*. Urbana: University of Illinois Press, 1997.

Harkness, Georgia Elma. *The Church and the Immigrant*. New York: George H. Doran Company, 1921.

Harrington, F. C. "Questions and Answers on the W.P.A." Washington, DC: U.S. Government Printing Office, 1939.

Hauser, Philip Morris, Evelyn Mae Kitagawa, Louis Wirth, and Chicago Community Inventory. *Local Community Fact Book for Chicago, 1950*. Chicago: Chicago Community Inventory, University of Chicago, 1953.

Hernandez-Fujigaki, Jorge. "Mexican Steelworkers and the United Steelworkers of America in the Midwest: The Inland Steel Experience (1936–1976)." Ph.D. diss., University of Chicago, 1991.

Hill, Jeff. *Sport, Leisure, and Culture in Twentieth-Century Britain*. New York: Palgrave, 2002.

Hines, Walker D. *War History of American Railroads*, New Haven: Yale University Press, 1928.

Hobsbawm, E. J., and T. O. Ranger. *The Invention of Tradition*. Cambridge: Cambridge University Press, 1983.

Hoerder, Dirk. *American Labor and Immigration History, 1877–1920s: Recent European Research*. Urbana: University of Illinois Press, 1983.

———. *"Struggle a Hard Battle": Essays on Working-Class Immigrants*. DeKalb: Northern Illinois University Press, 1986.

Hoffman, Abraham. *Unwanted Mexican Americans in the Great Depression: Repatriation Pressures, 1929–1939*. Tucson: University of Arizona Press, 1974.

Hondagneu-Sotelo, Pierrette. *Gendered Transitions: Mexican Experiences of Immigration*. Berkeley: University of California Press, 1994.
Honey, Michael K. *Black Workers Remember: An Oral History of Segregation, Unionism, and the Freedom Struggle*. Berkeley: University of California Press, 1999.
———. *Southern Labor and Black Civil Rights: Organizing Memphis Workers*. Urbana: University of Illinois Press, 1993.
Horowitz, Roger. *Negro and White, Unite and Fight! A Social History of Industrial Unionism in Meatpacking, 1930–90*. Urbana: University of Illinois Press, 1997.
Hughes, Elizabeth Ann. "Living Conditions for Small Wage Earners in Chicago." Chicago: City of Chicago Department of Public Welfare, 1925.
———, and Francelia Stuenkel. *The Social Service Exchange in Chicago*. Chicago: University of Chicago Press, 1929.
Humphrey, Norman D. "The Detroit Mexican Immigrant and Naturalization." *Social Forces* 22, no. 3 (1944).
———. "The Migration and Settlement of Detroit Mexicans," *Economic Geography* 19, no. 4 (1943).
Iber, Jorge, and Samuel O. Regalado, eds. *Mexican Americans in Sports: A Reader on Athletics and Barrio Life*. College Station: Texas A&M University Press, 2007.
Innis-Jiménez, Michael D. "Persisting in the Shadow of Steel : Community Formation and Survival in Mexican South Chicago, 1919–1939." Ph.D. diss., University of Iowa, 2006.
Jablonsky, Thomas J. *Pride in the Jungle : Community and Everyday Life in Back of the Yards Chicago*. Baltimore: Johns Hopkins University Press, 1993.
Jacobson, Matthew Frye. *Whiteness of a Different Color: European Immigrants and the Alchemy of Race*. Cambridge: Harvard University Press, 1998.
Jeter, Helen Rankin, Alfred Hamill, Chicago Commonwealth Club, and Chicago Regional Planning Association. *Trends of Population in the Region of Chicago*. Chicago: University of Chicago Press, 1927.
Jones, Anita Edgar. "Conditions Surrounding Mexicans in Chicago." M.A. thesis, University of Chicago, 1928.
Jones, Robert C., and Louis R. Wilson. *The Mexican in Chicago: The Racial and Nationality Groups of Chicago: Their Religious Faiths and Conditions*. Chicago: Comity Commission of the Chicago Church Federation, 1931.
Kelley, Robin D. G. *Hammer and Hoe: Alabama Communists During the Great Depression*. Chapel Hill: University of North Carolina Press, 1990.
———. *Race Rebels: Culture, Politics, and the Black Working Class*. New York: Free Press, 1994.
———. *Yo' Mama's Disfunktional! Fighting the Culture Wars in Urban America*. Boston: Beacon Press, 1997.
Kennedy, David M. *Over Here: The First World War and American Society*. New York: Oxford University Press, 1980.
Kerr, Louise Año Nuevo. "The Chicano Experience in Chicago, 1920–1970." Ph.D. diss., University of Illinois at Chicago Circle, 1976.
Kessler-Harris, Alice. *Out to Work: A History of Wage-Earning Women in the United States*. New York: Oxford University Press, 1982.
Kijewski, Marcia, David Brosch, and Robert Bulanda. *The Historical Development of Three Chicago Millgates: South Chicago, East Side, South Deering*. Chicago: Illinois Labor History Society, ca. 1973.

Kleinberg, S. J. *The Shadow of the Mills: Working-Class Families in Pittsburgh, 1870–1907*. Pittsburgh: University of Pittsburgh Press, 1989.

Kok, Jan. *Rebellious Families: Household Strategies and Collective Action in the Nineteenth and Twentieth Centuries*. New York: Berghahn Books, 2002.

LaGrand, James B. *Indian Metropolis: Native Americans in Chicago, 1945–75*. Urbana: University of Illinois Press, 2002.

Lane, James B., and Edward J. Escobar. *Forging a Community: The Latino Experience in Northwest Indiana, 1919–1975*. Chicago: Cattails Press, 1987.

Lang, Richard O., and Charles Shelton Newcomb. Census Commission. *Census Data of the City of Chicago, 1934*. Chicago: University of Chicago Press, 1934.

Lipsitz, George. *Time Passages: Collective Memory and American Popular Culture*. Minneapolis: University of Minnesota Press, 1990.

Lissak, Rivka Shpak. *Pluralism & Progressives: Hull House and the New Immigrants, 1890–1919*. Chicago: University of Chicago Press, 1989.

Martin, Phyllis. *Leisure and Society in Colonial Brazzaville*. New York: Cambridge University Press, 1995.

Massey, Douglas S. *Return to Aztlan: The Social Process of International Migration from Western Mexico*. Berkeley: University of California Press, 1987.

Mathewson, Stanley Bernard, Arthur Ernest Morgan, William Morris Leiserson, and Henry S. Dennison. *Restriction of Output among Unorganized Workers*. New York: Viking Press, 1931.

McCarthy, Malachy R. "Which Christ Came to Chicago: Catholic and Protestant Programs to Evangelize, Socialize and Americanze the Mexican Immigrant, 1900–1940." Ph.D. diss., Loyola University of Chicago, 2002.

McCombs, Vernon Monroe. *From over the Border: A Study of the Mexicans in the United States*. New York: Council of Women for Home Missions and Missionary Education Movement, 1925.

McWilliams, Carey. *North from Mexico: The Spanish-Speaking People of the United States*. New York: Greenwood Press, 1968.

Meyerowitz, Joanne J. *Women Adrift: Independent Wage Earners in Chicago, 1880–1930*. Chicago: University of Chicago Press, 1988.

Mohl, Raymond A. *The Making of Urban America*. Wilmington, DE: Scholarly Resources, 1988.

Monroy, Douglas. *Rebirth: Mexican Los Angeles from the Great Migration to the Great Depression*. Berkeley: University of California Press, 1999.

Montejano, David. *Anglos and Mexicans in the Making of Texas, 1836–1986*. Austin: University of Texas Press, 1987.

Montoya, Ramón Alejandro. *La Experiencia Potosina En Chicago*. San Luis Potosí, S.L.P., México: El Colegio de San Luis, 1997.

Moore, Powell A. *The Calumet Region; Indiana's Last Frontier*. Indianapolis: Indiana Historical Bureau, 1959.

Mormino, Gary Ross. "The Playing Fields of St. Louis: Italian Immigrants and Sports, 1925–1941." *Journal of Sports History* 9, no. 2 (Summer 1982): 5–19.

Nelson, Bruce. *Workers on the Waterfront: Seamen, Longshoremen, and Unionism in the 1930s*. Urbana: University of Illinois Press, 1988.

Nevins, Joseph. *Operation Gatekeeper: The Rise of The "Illegal Alien" and the Making of the U.S.-Mexico Boundary*. New York: Routledge, 2002.

Ngai, Mae M. *Impossible Subjects: Illegal Aliens and the Making of Modern America*. Princeton: Princeton University Press, 2004.

Odem, Mary E. *Delinquent Daughters: Protecting and Policing Adolescent Female Sexuality in the United States, 1885–1920*. Chapel Hill: University of North Carolina Press, 1995.

Orsi, Robert A. *Thank You, St. Jude: Women's Devotion to the Patron Saint of Hopeless Causes*. New Haven: Yale University Press, 1996.

Pacyga, Dominic A. *Polish Immigrants and Industrial Chicago: Workers on the South Side, 1880–1922*. Columbus: Ohio State University Press, 1991.

———., and Ellen Skerrett. *Chicago, City of Neighborhoods: Histories & Tours*. Chicago: Loyola University Press, 1986.

Pallares, Amalia, and Nilda Flores-González, eds. *¡Marcha! Latino Chicago and the Immigrant Rights Movement*. Urbana: University of Illinois Press, 2010.

Pascoe, Peggy. *What Comes Naturally: Miscegenation Law and the Making of Race in America*. New York: Oxford University Press, 2009.

Peck, Gunther. *Reinventing Free Labor: Padrones and Immigrant Workers in the North American West, 1880–1930*. New York: Cambridge University Press, 2000.

Phillips, Kimberley L. *Alabamanorth: African-American Migrants, Community, and Working-Class Activism in Cleveland, 1915–45*. Urbana: University of Illinois Press, 1999.

Roediger, David R. *Colored White: Transcending the Racial Past*. Berkeley: University of California Press, 2002.

———. *The Wages of Whiteness: Race and the Making of the American Working Class*. Rev. ed. New York: Verso, 1999.

———. *Working toward Whitenes: How America's Immigrants Became White: The Strange Journey from Ellis Island to the Suburbs*. New York: Basic Books, 2005.

Romo, Ricardo. *East Los Angeles: History of a Barrio*. Austin: University of Texas Press, 1983.

Rosaldo, Renato. *Assimilation Revisited.* Working Paper Series. Stanford: Stanford Center for Chicano Research, Stanford University, 1985.

Rosales, Francisco Arturo, and Daniel T. Simon. "Mexican Immigrant Experience Int He Urban Midwest: East Chicago, Indiana, 1919–1945." *Indiana Magazine of History* 77, no. 4 (December 1981): 333–57.

Rosales, F., Arturo. *Pobre Raza! Violence, Justice and Mobilization among Mexico Lindo Immigrants, 1900–1936*. Austin: University of Texas Press, 1999.

Rosales, Francisco Arturo, "Mexican Immigration to the Urban Midwest During the 1920s." Ph.D. diss., Indiana University, 1978.

Rosas, Ana E. "Breaking the Silence: Mexican Children and Women's Confrontation of Bracero Family Separation, 1942–64. *Gender & History* 23, no. 2 (August 2011): 382–400.

Rosenzweig, Roy. *Eight Hours for What We Will: Workers and Leisure in an Industrial City, 1870–1920*. New York: Cambridge University Press, 1983.

Ruíz, Vicki. *Cannery Women, Cannery Lives: Mexican Women, Unionization, and the California Food Processing Industry, 1930–1950*. Albuquerque: University of New Mexico Press, 1987.

———. *From Out of the Shadows: Mexican Women in Twentieth-Century America*. New York: Oxford University Press, 1998.

Rumbaut, Rubén G., and Alejandro Portes. *Ethnicities: Children of Immigrants in America*. Berkeley: University of California Press, 2001.

Sáenz, Moisés, Herbert Ingram Priestley, and Norman Wait Harris Memorial Foundation. *Some Mexican Problems: Lectures on the Harris Foundation 1926*. Chicago: University of Chicago Press, 1926.

Sánchez, George Isidore, and Stanford Center for Chicano Research. *Go after the Women: Americanization and the Mexican Immigrant Woman, 1915–1929*. Working Paper Series. Stanford: Stanford Center for Chicano Research, Stanford University, 1984.

Sánchez, George J. *Becoming Mexican American: Ethnicity, Culture, and Identity in Chicano Los Angeles, 1900–1945*. New York: Oxford University Press, 1993.

Schwalm, Leslie A. *A Hard Fight for We: Women's Transition from Slavery to Freedom in South Carolina*. Urbana: University of Illinois Press, 1997.

Scott, James C. *Weapons of the Weak: Everyday Forms of Peasant Resistance*. New Haven: Yale University Press, 1985.

Scott, James C., and Benedict J. Kerkvliet. *Everyday Forms of Peasant Resistance in South-East Asia*. London: Frank Cass, 1986.

Sepúlveda, Ciro. "La Colonia Del Harbor: A History of Mexicanos in East Chicago, Indiana 1919–1932." Ph.D. diss., Notre Dame University, 1976.

Shanas, Ethel, E. W. Burgess, Catherine E. Dunning, and Chicago Recreation Commission. *Recreation and Delinquency: A Study of Five Selected Chicago Communities*. Chicago: The Chicago Recreation Commission, 1942.

Shaw, Clifford Robe, Frederick McClure Zorbaugh, Leonard S. Cottrell, and Henry Donald McKay. *Delinquency Areas: A Study of the Geographic Distribution of School Truants, Juvenile Delinquents, and Adult Offenders in Chicago*. Chicago: University of Chicago Press, 1929.

Sheridan, Thomas E. *Los Tucsonenses: The Mexican Community in Tucson, 1854–1941*. Tucson: University of Arizona Press, 1986.

Smith, J. Douglas. *Managing White Supremacy: Race, Politics, and Citizenship in Jim Crow Virginia*. Chapel Hill: University of North Carolina Press, 2002.

Sugrue, Thomas J. *The Origins of the Urban Crisis: Race and Inequality in Postwar Detroit*. Princeton: Princeton University Press, 1996.

Taylor, Paul Schuster. *Mexican Labor in the United States: Chicago and the Calumet Region*, University of California Publications in Economics, vol. 7, no. 2. Berkeley: University of California Press, 1932.

———. *A Spanish-Mexican Peasant Community: Arandas in Jalisco*. Berkeley: University of California Press, 1933.

Thernstrom, Stephan. *Harvard Encyclopedia of American Ethnic Groups*. Cambridge: Belknap Press of Harvard University, 1980.

———. *The Other Bostonians: Poverty and Progress in the American Metropolis, 1880–1970*. Cambridge: Harvard University Press, 1973.

———. *Poverty and Progress: Social Mobility in a Nineteenth Century City*. Cambridge: Harvard University Press, 1964.

Trotter, Joe William. *Black Milwaukee: The Making of an Industrial Proletariat, 1915–45*. Urbana: University of Illinois Press, 1984.

Tuttle, William M. *Race Riot: Chicago in the Red Summer of 1919*. New York: Atheneum, 1970.

Valdés, Dennis Nodín. *Al Norte: Agricultural Workers in the Great Lakes Region, 1917–1970*. Austin: University of Texas, 1991.

———. *Barrios Norteños: St. Paul and Midwestern Mexican Communities in the Twentieth Century*. Austin: University of Texas Press, 2000.

———. *El Pueblo Mexicano En Detroit Y Michigan: A Social History*. Detroit: Wayne State University, College of Education, 1982.

Vargas, Zaragosa. *Proletarians of the North: A History of Mexican Industrial Workers in Detroit and the Midwest, 1917–1933*. Berkeley: University of California Press, 1993.
Warne, Colston E. *The Steel Strike of 1919*. Boston: Heath, 1963.
Warren, Kenneth. *Big Steel: The First Century of the United States Steel Corporation*. Pittsburgh: University of Pittsburgh Press, 2001.
Washington, Sylvia Hood. *Packing Them In: An Archaeology of Environmental Racism in Chicago, 1865–1954*. Lanham, MD: Lexington Books, 2005.
Weber, David J. *Foreigners in Their Native Land: Historical Roots of the Mexican Americans*. Albuquerque,: University of New Mexico Press, 1973.
Whitaker, Matthew C. *Race Work: The Rise of Civil Rights in the Urban West*. Lincoln: University of Nebraska Press, 2005.
Wilson, Tamar Diana. "Theoretical Approaches to Mexican Wage Labor Migration." *Latin American Persepctives* 20, no. 3 (1993): 98–129.
Wirth, Louis, Eleanor Harriet Bernert, and University of Chicago. *Local Community Fact Book of Chicago*. Chicago: University of Chicago Press, 1949.
Wirth, Louis, Eleanor Harriet Sheldon, and Chicago Community Inventory. *Local Community Fact Book of Chicago*. Chicago: University of Chicago Press, 1949.
Zamora, Emilio. *The World of the Mexican Worker in Texas*. College Station: Texas A & M University Press, 1993.
Zavella, Patricia. *Women's Work and Chicano Families: Cannery Workers of the Santa Clara Valley*. Ithaca: Cornell University Press, 1987.

INDEX

ABOUT THE AUTHOR

A native of Laredo, Texas, Michael Innis-Jiménez is Assistant Professor in the Department of American Studies at the University of Alabama. He lives in Tuscaloosa, where he is working on his next book on Latino/a immigration to the American South.